Thomas Kuhn
In the Light of Reason

By

Brian Maricle

Thomas Kuhn: In The Light of Reason

Copyright © 2008 by Brian Maricle

All Rights Reserved.

No Part of this book may be reproduced in any manner whatsoever without written permission of the author except in brief quotations embodied in critical articles or reviews.

All quotations from Bernard Lonergan's book *Insight* have been reprinted with permission of the publisher.

First Printing 2008

Second Edition published 6/17/2008

ISBN - 978-0-9747930-0-9

Those who put together difficult, obscure, involved, ambiguous discourses ... often ... want to conceal from themselves and others that they actually have nothing to say.

- Schopenhauer

Preface to the Second Edition

What is science? One could attempt to answer this question in different ways. For instance, one might answer this question by simply describing the behavior of scientists. However such a description may or may not clarify the nature of *successful* science. The first edition discusses the nature of successful science but this is such a foundational point that it needed to be brought into sharper focus in the first chapter of the second edition.

Another fundamental point needed clarification which is this: The principle of causality is the basis of science. If there were no casual relations, then science would not be possible. One cannot build rockets, cars and telephones based on purely random events. If there were no casual relations, then Quantum Mechanics could never have been used to build high quality precision machines. Nevertheless, in popular literature, Quantum Mechanics is often presented as evidence that law of cause and effect no longer plays an important role in science which has given credence to irrational characterizations of modern science. This subject is, therefore, addressed in the last chapter of the second edition.

Table of Contents

Acknowledgements.. 1

A Simple Road Map... 3

I   *The Light of Reason*.. 5

II   *Discovery*

        1. The Pattern of Rational Intelligence............. 15

        2. Empirical Consciousness............................. 18

        3. Intelligent Consciousness............................ 26

        4. Rational Consciousness............................... 46

        5. The Nature of Scientific Discovery............... 57

III *Knowledge*

        6. The Nature of Scientific Knowledge............. 67

        7. Einstein Builds on Newton's Theory 73

IV *Loss of Confidence in Reason*.. 91

Bibliography.. 103

Acknowledgements

I am most grateful to Bernard Lonergan, the celebrated Jesuit philosopher, who has been named the Thomas Aquinas of the 20th Century and without whom my efforts to write a book about the nature of science would have been in vain. This entire work may be regarded as an effort to confront Thomas Kuhn's philosophy of science with the philosophic rationality of Bernard Lonergan. In particular, I have been guided by Lonergan's ideas on the important function of clues in the search for knowledge, the foundational significance of measurement to modern science and the nature of rational knowledge.

I am also indebted to the brilliant Australian philosopher, David Stove for his psychological perceptiveness in identifying the historic origin of our modern uncertainty about the truth of scientific claims.

I am most grateful to all my friends and family for their love and support. In particular, I am thankful to Karin Nelson for providing invaluable feedback with regard to editing this manuscript as well her insightful and much needed assistance in helping me significantly improve my skills as a writer.

I am also thankful to Douglas Lee for generously sharing with me his exceptionally clear understanding of mathematical and scientific matters.

I am grateful to my brother Chris for editing this work and especially for his valuable help in the difficult task of organizing this material.

I am grateful to Jack Lincoln for many spirited and inspiring discussions regarding the nature of science and truth.

I am grateful to Scott Furman for reading this material and providing helpful feedback.

I am thankful to John Tellison for his feedback regarding these ideas and especially for his skilled and generous help in creating the cover.

A Simple Road Map

This book is divided into the following four sections:

I. The Light of Reason

The first section explains the two main points of this book:

1) Thomas Kuhn distorted the rational nature of science.
2) This distortion significantly contributed toward raising doubts about the importance of reason.

II. Discovery

The second section establishes the rational nature of scientific discovery that Kuhn neglected.

III. Knowledge

The third section shows how Kuhn excluded the sense in which modern science is a rational and unchanging knowledge of the physical world.

IV. Loss of Confidence in Reason

The forth section tells the psychological story of how Kuhn's work on the philosophy of science undermined confidence in the effectiveness of rational thought.

The Light of Reason

One of the most influential books ever written on the nature of science is *The Structure of Scientific Revolutions*, by Thomas Kuhn. According to the *Arts and Humanities Index*, this book was the most cited single work of the 20th Century.[1] Random House and the *National Review* both include this book as one of the one hundred best non-fiction books of the century.

Kuhn's book, published in 1962, challenged the traditional view that science continually advances—each new discovery is a step toward a more complete understanding of nature. Kuhn presented a different image of science. He argued that new scientific discoveries are not like pieces of a jigsaw puzzle that fit neatly together to form a single coherent scientific picture of the world. Rather, the picture itself periodically changes. Kuhn called such fundamental changes *paradigms shifts*, which can be compared to the shifting perspectives of a kaleidoscope where: no perspective is the "correct" perspective; no perspective follows developmentally from any other; and no perspective tells one anything about the *true* nature of reality. In order to illustrate this point, Kuhn claimed that Einstein's theoretical concepts fundamentally replaced the scientific concepts of Newton, but, as we shall see, this claim is untrue; Einstein, himself, believed unequivocally that he built on the scientific achievement of Sir Isaac Newton.

Kuhn's new vision of scientific truth as periodically subject to fundamental changes had the effect of significantly raising doubts about the value of reason, an effect that may be explained in the following way: Reason would be useless in a universe where

---

[1] *Arts and Humanities Citation Index*, second semiannual

the basic nature of reality changed from one moment to the next. For instance, suppose I infer from past experience that I will fall if I should step out of an open window. However, my inference would be mistaken if the law of gravity ceased to function, in which case I might simply float away like a cloud. Reason is only effective insofar as things remain constant. By making scientific truth appear changeable, Kuhn made the world appear less rational and the power of reason less effective.

This book would have no value if reason were unimportant; therefore, the value of reason must be clear at the onset. With regard to practical matters, the value of reason is self-evident and universal. For human beings, the ability to reason—to be rational, not to rationalize—is an important resource for making useful decisions. Everyday examples abound. When we buy a car, the intelligent thing to do is not to buy the first car that we see, but to judge which car is the best purchase based on the most reliable information. Reason is required for every facet of daily living. Building homes, providing food and raising children all require good reasoning and sound judgment. Furthermore, all of modern business, law, mathematics, and technology, along with our entire system of justice, depend upon the use of reason. Philosophy, itself, without reason, would be little more than myth or speculation. Nevertheless, establishing the importance of reason to human life has been a long uphill battle. Consider the fate of Galileo, whose rational pursuit of scientific truth was such a threat to the Church that he was ordered to publicly renounce his scientific convictions or face the prospect of being tortured. Consider the fate of Socrates, whose rational pursuit of philosophical truth was seen as such a threat to the state that he was condemned to death.

It was the light of reason that guided Europe out of the dark ages and into the Enlightenment and the Age of Reason (approximately 1680–1780), which inspired among other things a rational pursuit of human rights. Subjecting political and religious authority to the demands of rational thought was the ideal that guided the writing of the Declaration of Independence and the United States Constitution.[2] Personal happiness and individual rights emerged as among the highest values of rational Enlightenment philosophy.

The astounding success of scientific reasoning during the Enlightenment provided powerful evidence that nature yields her secrets more readily to rational inquiry than to any other way of obtaining knowledge. For upon publication of Isaac Newton's famous book, *The Mathematical Principles of Natural Philosophy*, the world soon became acquainted with precise mathematical equations that could quantitatively predict the movements of heavenly bodies, including earthbound phenomena such the behavior of the tides. The world has seen nothing like this explosion of scientific understanding before or since the time of Newton. With the overwhelming success of Newtonian science, confidence in the power of rational thought soared to an all-time high.

And yet, today, rationality no longer commands the respect it once did and in some circles is even regarded with scorn.[3] This strange fact is to be explained, in part, by the rise of

---

[2] Reason was so highly held to be the key to a wise and just life that Kant wrote: "Have courage to use your own reason—that is the motto of the Enlightenment."

[3] *Return to Reason*, Stephen Toulmin, "Eighty or ninety years ago, scholars and critics, as much as natural scientists, shared a common confidence in their

scientific irrationality—modern views on the nature of science that minimize the significance of reason. For if one supposes that science is not fundamentally rational, it follows that the remarkable effectiveness of modern science is not essentially the product of rationality, and if so, then the significance of reason in western culture has been greatly exaggerated and the corresponding loss of confidence in the power of human reason fully justified. This is the psychological legacy of Thomas Kuhn, who wrote about science in such a way as to greatly reduce the meaning and significance of scientific rationality. In doing so, he not only clouded our understanding of science, but he also cast a shadow of doubt on the fundamental importance of reason. The length of that shadow can be measured by the success of his book, *The Structure of Scientific Revolutions*.

A few clarifying remarks may be helpful. The overall strategy of this book is to demonstrate the rationality of science by showing that *science is nothing more than an expansion of rational common sense*. Therefore, although this book embraces a distinct philosophical viewpoint, one must not expect a philosophical defense of that view. This book appeals to the common sense of the reader; consequently, the reader, alone, is the sole and absolute authority on whether or not the appeal succeeds.

---

established procedures. The term "'scientific method'" embraced, for them, all the methods of observation, deduction, generalization, and the rest that had been found appropriate to the problems and issues preoccupying those subjects. How little of that confidence remains today. Among some humanists, the phrase "'scientific method'" is even pronounced with a sarcastic tone; and one even hears it argued that the concept of rationality itself is no more than a by-product of Western or Eurocentric ways of thinking."

It may be helpful to state what this book is not about. Although paradigm shifts are discussed within these pages, this book is not principally about paradigm shifts or the extent to which they may or may not occur. The aim of this book is to demonstrate the implausibility of Thomas Kuhn's scientific irrationality by showing that reason is the foundation of science. And, in doing so, perhaps, to restore a measure of confidence in the high importance of reason that has been lost since the Enlightenment.

Discovery

Chapter 1

The Pattern of Rational Intelligence

Scientists are most effective when they follow the same pattern of investigation to be found in any rational inquiry which consists of three basic steps: (1) making observations; (2) raising questions and forming an understanding, then (3) checking to make sure the proposed understanding is correct. For example, suppose I *observe* smoke rising from my house and *inquire* into what's going on. After some thought, I reach the *understanding* that my house appears to be on fire. I look for evidence to see if I am right and find, to my great relief, that I am wrong and that my nephew has built a little fire in the back yard. I then *judge* that the evidence does not support the theory that my house is on fire. In the above scenario, I moved through three distinct levels of consciousness:

(1) **Empirical Consciousness**—the awareness of sense perception (seeing smoke rising from my house).
(2) **Intelligent Consciousness**—asking questions and reaching understanding (wondering why smoke is rising from my house and forming the understanding that my house could be on fire).
(3) **Rational Consciousness**—Judging if the understanding is supported by evidence (deciding that there is no good reason to believe that my house is on fire).

The above three stages constitute the core of Bernard Lonergan's brilliant theory of knowledge.[4] His position is that acquiring rational knowledge begins with Empirical Consciousness, an awareness of the world disclosed through the senses, which includes sights, sounds, shapes and colors. Desiring to understand the data of sense experience emerges as questions, ushering us into Intelligent Consciousness, a stage characterized by curiosity, seeking insight and acquiring understanding. Asking questions such as, "Am I right ?" or "Could I be mistaken?" moves us into Rational Consciousness, which consists of critically reflecting, carefully weighing evidence and double checking assumptions, a stage that culminates by answering "yes" or "no" to the question "Is it so?"

Part of the originality of Lonergan's theory of knowledge is the fundamental significance he accords to the *desire to know*, which, like a powerful motor, drives the whole process of acquiring rational knowledge. By urging us to understand the world that we apprehend through sense experience, the desire to know raises us from Empirical Consciousness to Intelligent Consciousness. By urging us to understand *correctly*, the same desire ushers us from Intelligent Consciousness to Rational Consciousness.

---

[4] Bernard Lonergan, Insight pages 298-299.

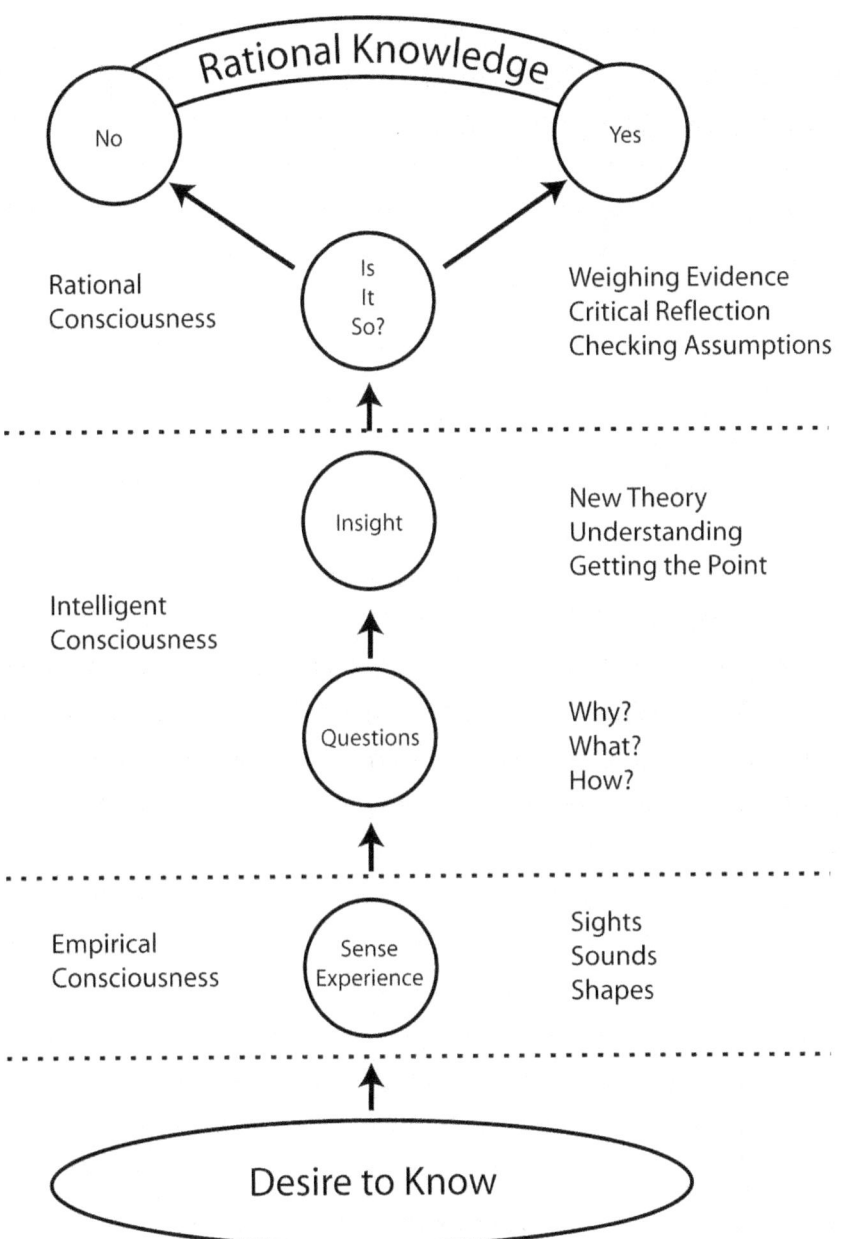

With a bow toward Lonergan, let us refer to these three stages as the Pattern of Rational Intelligence. These steps constitute a *pattern* insofar as they are a recurrent theme in everyday life, the pattern is *intelligent* for it depends upon insight, and the pattern is *rational* for it demands sufficient reason to affirm a proposed understanding as correct.

One purpose of this work is to demonstrate that science is not a strange, mysterious, unfamiliar kind of thinking, but one that follows the same basic steps to be found in everyday life. A good scientist, like every good detective, (1) makes observations, (2) seeks insight, forms an understanding, and then, (3) based on available evidence, judges whether or not such an understanding is correct.

Just as we study great artists to improve our comprehension of Art, we shall study a great scientist to improve our understanding of science. Therefore, to illustrate how successful science follows the Pattern of Rational Intelligence; we shall turn to one whom many regard as the Founder of Modern Science, Galileo[5].

---

[5] Einstein declared Galileo to be the Father of Modern Science, *Ideas and Opinions*, page 271., Richard Feynman held that, prior to Galileo, theories about the nature of motion were speculative rather than scientific, *The Feynman Lectures on Physics*, Vol. 1, page 5–1., Stephen Hawking said that modern science owes it origins to Galileo more than any other person, *A Brief History of Time*, page 194.

Chapter 2

Empirical Consciousness

Scientific inquiry yields the most impressive results when it follows the Pattern of Rational Intelligence:

(1) Empirical Consciousness (awareness of sense perceptions)
(2) Intelligent Consciousness (asking questions and forming an understanding)
(3) Rational Consciousness (determining correctness by weighing evidence)

Beginning with Empirical Consciousness, we will consider what exactly scientists are out to understand and then what motivates them to do so.

Sense Experience: The Object of Modern Scientific Inquiry

Part of Galileo's genius was that he recognized the scientific significance of observation long before it was generally appreciated. Today, the idea that observation is an effective way to learn about physical reality is practically self-evident. We, of course, have the benefit of four centuries of scientific experience repeatedly trumpeting the value of observation, but this was not always so obvious. For instance, Plato, one of the greatest philosophers of ancient Greece, believed that the realm of pure ideas was more real and more perfect than the fleeting world of sense perception, which was more prone to illusion and error.

Therefore, a reliable knowledge of reality was to be obtained turning away from the misleading world of sense perception and towards a pure contemplation of true and unchanging ideas.[6]

Galileo had not yet impressed upon the world that *science begins with observing the concrete world of sensory experience*—that paying attention to what *actually* happens via sense-experience is a better way to learn about nature than imagining what *ought* to happen. For instance, Ptolemy, the famous astronomer of ancient Greece, believed that the planets must move in perfect circles because a circle is a perfect shape. A scientist in the Galilean spirit, on the other hand, would tell you that if you wanted to know the shape of planetary orbits, don't start by imagining how the planets *ought* to move and then look for evidence to support what you imagine. Instead, *observe* how they *actually* move, try to understand what you are observing, and treat observation as the final authority on whether you understand correctly or not. This extremely important point constitutes a major contribution to science. It was because Galileo emphasized learning about physical reality through sensory experience that Einstein bestowed upon him the title of "Father of Modern Science":

> "Pure logical thinking cannot yield us any knowledge of the empirical world; all knowledge of reality starts from experience and ends in it…

---

[6] This idea is perhaps best expressed by Plato's famous Cave allegory where prisoners are chained in a cave from birth. All they can ever see is the back wall of the cave and are unable to turn their heads. There is a fire behind them where puppeteers hold up puppets that cast shadows on the wall. The prisoners are unable to the see the real objects (the puppets) and mistake the puppet-shadows on the wall for reality.

Because Galileo saw this, and particularly because he drummed it into the scientific world, he is the father of modern physics—indeed, of modern science altogether."[7]

The Importance of Experiment

As Einstein indicates, science begins and ends with sensory experience; it begins with physical observations and concludes with observing the results of physical experiments. The scientific value of experiment is so deeply ingrained into our thinking that it's difficult to comprehend how any educated, intelligent person could fail to see the significance of it. But Aristotle was neither uneducated nor unintelligent, and yet experimental evidence played no significant part of his investigations into nature. The ancient Greeks were no less intelligent than we are today; they simply had little or no experience with the astonishing effectiveness of experimentation in acquiring a trustworthy knowledge of the physical world. Moreover, as we mentioned earlier, there was sometimes even a distinct bias against sense experience. This was the case with Plato, who believed that knowledge of reality could be obtained by a pure and highly rarified contemplation—unlike Aristotle, who had some clue as to the importance of observation and was himself remarkably observant. For instance, Aristotle discerned that the earth was round based on the observation that when ships are sailing toward the horizon, they gradually disappear (as though they were sinking into the sea). Aristotle reasoned that the appearance of ships sinking into the ocean could be explained by supposing

---

[7] Albert Einstein, *Ideas and Opinions*, page 271.

that the earth had a round shape. He also inferred the roundness of the earth from the curved shadow of the earth that falls on the moon during eclipses. Still, he did not recognize the importance of experimentation in scientific matters.

It may be worth adding that Aristotle stood somewhat between Galileo and Plato with regards to science. Aristotle realized to a great extent the value of observation and reason. For instance, he made many striking observations with regard to biology and founded western logic—no small achievement. But his appreciation of the importance of sense experience did not lead him to check his assumptions by means of physical experiments, nor did he make any measurements, which, as we shall see, are the signs of modern science.

Wonder—The Source of All Philosophy and Science

The objective of science is to understand the data of sensory experience. But what motivates scientific discovery? Consider Kuhn's position on this subject.

The birth of a fundamentally new scientific theory, Kuhn tells us, begins with anomaly—the case where nature does not behave as expected.[8] The repeated experience of nature not behaving as expected can make scientists insecure,[9] a fear that

---

[8] Thomas Kuhn, *The Structure of Scientific Revolutions*, "Discovery commences with the awareness of anomaly," page 52.
[9] Thomas Kuhn, *The Structure of Scientific Revolutions*, "The emergence of new theories is generally preceded by a period of pronounced professional insecurity. As one might expect, that insecurity is generated by the persistent failure of the puzzles of normal science to come out as they should. Failure of existing rules is the prelude to a search for new ones,"" page 67-68.

can become so great that some men actually abandon science because they can't handle the pressure.[10] When the level of professional insecurity reaches a certain point, the scientific community is officially in a state of crisis. This state of mind is held to be necessary for new discoveries to take place.

Kuhn correctly observed that a fundamental divergence between observation and theoretical expectation can cause insecurity—there are documented cases of this. And yet, is such fear a *necessary* pre-condition[11] for discovery? Fear is not commonly the driving force behind exploration of *any* kind. One does not explore the depths of the ocean or high mountains because one is afraid, but rather out of a desire to know and experience new things. Surely, the quest for scientific knowledge can be as adventurous as any exploration.

Aristotle said that wonder is the source of all science and philosophy.[12] Einstein made a similar observation:

> "There exists a passion for comprehension, just as there exists a passion for music. That passion is rather common in children, but gets lost in most people later on. Without this passion, there would be neither mathematics nor natural science."[13]

---

[10] Thomas Kuhn, *The Structure of Scientific Revolutions*, "Some men have undoubtedly been driven to desert science because of the inability to tolerate crisis," page 78-79.
[11] Thomas Kuhn, *The Structure of Scientific Revolutions*, "Let us then assume that crises are a necessary precondition for the emergence of novel theories," page 77.
[12] Bernard Lonergan, *Insight*, page 34.
[13] From the article "On the Generalized Theory of Gravitation," *Scientific American*, Vol. 182, No. 4, April, 1950.

Good scientists are motivated by a desire to know. Great scientists are passionate about the knowledge they seek and possess a tremendous capacity for understanding it. Wonder was Newton's general feeling about the world:

> "I do not know what I may appear to the world, but to myself I seem to have been only like a boy playing on the sea shore, and diverting myself in now and then finding a smoother pebble or a prettier shell than ordinary, whilst the great ocean of truth lay all undiscovered before me."[14]

Was Newton's motive for discovery anxiety or a sense of wonder? Consider what Einstein said:

> "The main source of all technological achievements is the divine curiosity and playful drive of the tinkering and thoughtful researcher, as much as it is the creative imagination of the inventor."[15]

> "It is my inner conviction that the development of science seeks in the main to satisfy the longing for pure knowledge."[16]

> "The mainspring of scientific thought is not an external goal toward which one must strive, but the pleasure of thinking."[17]

---

[14] From Brewster's biography, op. cit., Vol. II, ch. 27.
[15] August 22, 1930, in a radio broadcast in Berlin.
[16] 1920, in Moszkowski, *Conversations with Einstein*, page 73.

The motives that lead to scientific discovery, according to Einstein, are not fear or anxiety but divine curiosity, the longing for pure knowledge and the pleasure of thinking. Historian of science John Gribbon put it this way:

> "With very few exceptions, scientists throughout history have plied their craft not through a lust for glory or material reward, but in order to satisfy their own curiosity about the way the world works."[18]

James Clerk Maxwell, one of the greatest scientists of the 20th century, has this to say about the nature of scientific discovery:

> "We, while following out the discoveries of the teachers of science, must experience in some degree the same desire to know and the same joy in arriving at knowledge which encouraged and animated them."[19]

Lonergan tells us just how passionate the desire for comprehension can be:

> "It can absorb a man. It can keep him for hours, day after day, year after year, in the narrow prison of

---

[17] to Heinrich Zangger, ca August 11, 1918, CPAE, Volume 8, Doc. 597.
[18] John Gribbon, *The Scientists*, page 613.
[19] Harman, P. M. (ed.) (1990) *The Scientific Letters and Papers of James Clerk Maxwell* Vol. I, Cambridge: Cambridge University Press.

his study or his laboratory. It can send him on dangerous voyages of exploration. It can withdraw him from other interests, other pursuits, other pleasures, other achievements. It can fill his waking thoughts, hide him from the world of ordinary affairs, invade the very fabric of his dreams. It can demand endless sacrifices that are made without regret though there is only the hope, never a certain promise, of success"[20]

The Special Theory of Relativity was not born of anxiety, but of a wonder and curiosity that Einstein experienced as a young man when he wondered what it would be like to ride a wave of light.[21]

Kuhn depicted discovery as anxiety-driven more than anything else. But according to Aristotle, Newton and Einstein, the true motives of a scientist are wonder, curiosity, the pleasure of thinking, and a passionate desire to comprehend the mysteries of the universe.

Summary

(1) The object of science is to comprehend the data of sense experience.
(2) Scientific understanding begins with a desire to know.

---

[20] Bernard Lonergan, *Insight*, page 28.
[21] Max Wertheimer, *Productive Thinking*, page 218.

Chapter 3

Intelligent Consciousness

Kuhn excluded the intrinsically rational nature of scientific inquiry from his account of it. To remedy this omission, we have been demonstrating how science follows the Pattern of Rational Intelligence, which proceeds as follows:

(1) Empirical Consciousness (awareness of sense perceptions)
(2) Intelligent Consciousness (asking questions and forming an understanding)
(3) Rational Consciousness (determining correctness by weighing evidence)

Scientific inquiry begins with a desire to understand the data of sensory experience. And yet, how does one get from inquiry to insight? When we want to understand something, what do we do? Do we ask questions and then make random guesses? Sometime we do, but according to Lonergan, the following conditions increase the likelihood of achieving insight:

1. A Desire to Know Undistorted by Personal Bias
2. A Capacity for Insight
3. Clues

A Desire to Know Undistorted by Personal Bias

Who among us has not discovered that it is difficult to be fair and objective about a matter where one's personal interests are at stake? This is why judges are not allowed to preside over cases where they have a vested personal interest, such as being related to the accused. Even if the judge in question sincerely tries to be fair, a personal interest in the matter makes it difficult to be objective. In this respect, scientists are not different from other people. Having a vested personal interest in whether or not a given theory is correct makes it difficult, if not impossible, to be objective. A scientist, like everyone else, achieves insight more readily when personal bias does not distort the desire to know. Lonergan describes a scientific frame of mind this way:

> "The interests and hopes, desires and fears, of ordinary living have to slip into a background. In their place the detached and disinterested exigencies of inquiring intelligence have to enter and assume control ... just as the woodsman, the craftsman, the artist, the expert in any field acquires a spontaneous perceptiveness lacking in other men, so too does the scientific observer."[22]

A Capacity for Insight

The achievement of understanding depends in part upon a capacity for insight. What is meant by Insight? It generally means to get the point, to see what's up or grasp what's at issue. It can be said that insight is present when you understand and absent when you do not. Lonergan observes:

> "...insight is the act that occurs frequently in the intelligent and rarely in the stupid"[23]

---

[22] Bernard Lonergan, *Insight*, page 97.
[23] Ibid., page 29.

Lonergan boldly asserts that the capacity for insight is the defining characteristic of human intelligence. Thus, a genius is one who has a capacity for insight into matters that escapes the comprehension of most, whereas a fool is one to whom the occurrence of insight is a rare experience.

Consider the nature of insight more closely. Lonergan tells us: "Insight comes as a release to the tension of inquiry."[24] He is saying that insights are not random events. Specifically, they are a response to *something*—a desire to know. This desire urges us to seek understanding and draws upon intellectual resources such as hunches, perceptions, imagination, reason, creativity, concepts, intuition and memory. There is no telling when or if the light of comprehension will dawn, but such understanding is generally preceded by a desire to know, even if the desired insight comes unexpectedly.[25]

Clues

An unbiased desire to know combined with a capacity for insight increases the likelihood that a search for understanding will succeed, and the chances are further improved if we can find clues[26] that point toward a solution.

What exactly is a clue? The meaning of the word comes from the Greek myth of Theseus and the Minotaur in which a vast

---

[24] Bernard Lonergan, *Insight*, page 28.
[25] Bernard Lonergan, *Insight*, page 28.
[26] Part of the originality of Lonergan's epistemology is the unique significance that he attaches to the important role that *clues* play in acquiring knowledge, Bernard Lonergan, Insight page 60,

labyrinth was built as a prison for the Minotaur, a creature that was half man and half bull. Theseus sought to kill the Minotaur, which unfortunately meant going into the labyrinth where there was little hope of ever finding the way out. Theseus was presented with a problem: he needed to discover a way of getting out of the Labyrinth. He solved his dilemma by unrolling a ball of thread as he went and escaped by following the thread back the way he had come.

The word "clue" comes from the Greek word *clew*, which literally meant a ball of thread or yarn and has come to mean anything that leads to discovery. Consider some important characteristics of clues illustrated by Theseus:

(1) A clue serves to guide an inquiry. If one is lost in a labyrinth, following the thread guides one to the exit. Without the thread, one can only search at random for the way out.
(2) A clue connects the known to the unknown. At any given moment, Theseus knew that he was somewhere inside a vast labyrinth. The thread connected Theseus to what he did not know—the way out.
(3) Since a clue connects the known to the unknown, following a clue that leads to discovery extends what is known.

An Everyday Example of Clues

First, consider the function of clues in a criminal investigation. Suppose that someone was killed with a gun and the fingerprints have been found. The detective's inquiry begins with the desire to know who committed the murder. His desire is guided by a clue—the fingerprints, the part of the solution that is known. The detective is fortunate to have a clue, because without it, he can only guess. The fingerprints are the detective's thread in the labyrinth of searching for the murderer. In solving the crime, the detective extends what he knows—the murderer's fingerprints—to include what he does not know—the identity of the murderer.

Clues in Scientific Investigations

Now, let's consider an example from science: Galileo's discovery that a free fall is a constant acceleration. Not only does this discovery illustrate how clues function in a scientific investigation, this particular discovery marks the transition from pre-scientific thought to modern science.

In order to appreciate the significance of Galileo's discovery, let us first take a look at how Galileo's predecessors investigated the natural world. Prior to Galileo, Aristotle was held to be the supreme authority on scientific matters. According to Aristotle, there are four basic types of explanation: material, formal, efficient, and final. For example, consider how Aristotle might explain a house:

(1) **Material explanation**: a house is made of wood, metal, glass and cement.
(2) **Formal explanation**: a house is defined by four walls and a roof (or blueprints).
(3) **Efficient explanation**: a house is produced by construction workers.
(4) **Final explanation**: the purpose of a house is to provide a place where people can live.

Thus, an Aristotelian explanation of a free fall might distinguish between the different kinds of materials that fall, identify what produces a fall and seek to discover the final explanation for falling objects. For instance, Aristotle believed that the ultimate explanation for falling objects was that earthly objects are naturally drawn to the earth and that heavier materials fall more quickly than lighter materials. Galileo attempted none of these explanations.

A Clue from Mathematics

Aristotle, like Galileo, sought to understand the nature of a free fall. Both observed, inquired and reasoned. But Galileo took additional steps. In the first place, he proceeded with a clue from mathematics:

> "Galileo supposed that some correlation was to be found between the measurable aspects of falling bodies. Indeed, he began by showing the error in the ancient Aristotelian correlation that bodies fell according to their weight. Then he turned his attention to two measurable aspects imminent in

every fall: the body traverses a determinate distance; it does so in a determinate interval of time. By a series of experiments he provided himself with the requisite data and obtained the desired measurements. Then he discovered that the measurements would satisfy a general rule: the distance traversed is proportional to the time squared. It is a correlation that has been verified directly and indirectly for over four centuries.[27]"

The clue from mathematics was Galileo's thread in the labyrinth of understanding the nature of a free fall. He extended what he knew (mathematics) to include what he did not know (the mathematical formula for an object in free fall). Not only did the clue from mathematics guide him to a new discovery, the anticipation that natural phenomena can be mathematically understood has marked the scientific effort ever after. Let us, then, pause our discussion about the function of clues to consider the importance of mathematics to modern science.

Mathematics—A Defining Character of Modern Science

To say that math is an effective way to understand nature does not sound like an extraordinary thing to say—at least not in the 21st century. On the contrary, the usefulness of math to science is virtually self-evident. But this was not always true. In the 17th century, math was certainly not considered necessary to the study of nature. When Galileo sought a mathematical

---

[27] Bernard Lonergan, *Insight*, page 58.

understanding of a free fall, he was doing something *very* unusual.[28]

It is difficult to appreciate the excitement and interest Galileo must have felt at pursuing a mathematical understanding of nature—an idea we now take for granted. Norwood Russell Hanson put it this way:

"What then was discovery to Galileo? It was the perception of cohesive, mathematical structure within the buzzing detail of experience. For him, every falling coin, every windblown leaf, every new moon was a special kind of anomaly, an occasion for inquiry. Phenomena like these, familiar but not understood, were the windows through which the anatomy of the universe could be witnessed, if one but focused the appropriate mathematical lens."[29]

Galileo once said:

"The book of nature cannot be understood unless one learns to comprehend and to read the alphabet in which it is composed. It is written in the language of mathematics."[30]

Four centuries after Galileo, we can send people to moon, communicate instantly over great distances and possibly

---

[28] For a clear and interesting account of unusualness of Galileo's mathematical approach see *Morris Kline Mathematics and the Physical World*, Chapter 11.
[29] Morton F. Kaplan, *Homage to Galileo*, page 49.
[30] Stillman Drake, *Galileo*, page 170.

destroy the whole planet if we are not careful. Who would have thought that applying math to nature could lead to such swift and gigantic strides in understanding the world?

Galileo was a brilliant scientist, but his ideas about science did not occur in a vacuum; his desire to understand the physical world was to some extent guided by the mathematical and scientific insights from the past.

Galileo's Indebtedness to Archimedes and Euclid

Galileo's scientific hero from ancient Greece was Archimedes[31] (287–212 BC), considered by many to be the greatest mathematician and scientist of the ancient world. The exploits of Archimedes are legendary. He was a brilliant inventor who, among many other things, designed ingenious machines of war that were far ahead of their time. For example, according to one story, Archimedes is said to have prevented an attack from the sea by using a "burning glass," which reputedly reflected the light of the sun causing the enemy ships to burst into flame. It has also been reported that Archimedes designed a large crane-like device with a grappling hook that could be used to literally grab enemy ships, lift them out of the water and tip them over. He was a dramatic figure whose work with levers led him to proclaim: "Give me a place to stand on, and I will move the earth." He also made discoveries in mathematics, geometry and hydrostatics. And perhaps most importantly, Archimedes had hit upon one of the keys to modern science—he applied mathematical models to the

---

[31] William R. Sheaw, *Galileo's Intellectual Revolution*, Middle Period 1610–1632, page 1.

study of the physical world. No wonder that Galileo found him inspiring.

Another major influence on Galileo was Euclid,[32] who lived around 300 BCE. Euclid devised a system of geometry that was to be the *only* geometry for the next 2,000 years. According to Einstein, Euclid's brilliant display mathematical logic was of fundamental importance to western science:

> "We reverence ancient Greece as the cradle of western science. Here for the first time the world witnessed the miracle of a logical system, which proceeded step by step with such precision that every single one of its propositions was absolutely indubitable – I refer to Euclid's geometry. This admirable triumph of reasoning gave the human intellect the necessary confidence in itself for its subsequent achievements"[33]

Euclid's book *Elements* is one of the most influential books ever written, but he did not invent geometry; the geometrical knowledge contained in Euclid's book, was to a large extent known to the Egyptians who made use of geometric techniques in order to build the pyramids.

Euclid's book demonstrated the remarkable power of logical deduction. He logically deduced an entire system of geometrical theorems from just a few simple premises, thereby

---

[32] Stillman Drake, *Galileo: Pioneer Scientist*, page 213.
[33] From a lecture entitled *On the Methods of Theoretical Physics,* Oxford, June 10, 1933.

integrating geometry into a logical coherent system of thought that Einstein found to be so impressive. Euclid's model of mathematical reasoning is still the rational basis of modern scientific thought.

The discoveries of Galileo were not the outcome of chance. On the contrary, his scientific investigations were guided by clues from the past, especially by the celebrated works of Archimedes as well as the achievement of Euclid's geometry.[34]

Newton Enriches the Clue from Mathematics

Guided by the clue of mathematics, Sir Isaac Newton discovered mathematical relationships that could explain the tides and predict with great accuracy the positions of the planets and their moons and other heavenly bodies. Newton published his discoveries in a book called, *The Mathematical Principles of Natural Philosophy*. "Natural philosophy" meant the study of the physical world.

Newton enriched Galileo's mathematical clue in the following manner. In order to find mathematical relations that could be applied to nature, Newton developed techniques for calculating rates of change known as differential equations.

Since Newton's time, the use of differential equations has grown exponentially. It is no exaggeration to say that most of the

---

[34] "Galileo was particularly influenced by the model of Archimedean hydrostatics as well as Euclid's theory of proportions from Book V of *Euclid's Elements*," Paolo Palmieri. From an article entitled *The Cognitive Development of Galileo's Theory of Buoyancy,* Published online October 26, 2004, Springer-Verlag 2004.

laws of physics are (or can be) expressed in the form of differential equations.[35] In time, partial differential equations[36] evolved from differential equations. According to Einstein, such equations express the "...primary realities of physics."[37] Today when scientists study nature, they are guided by the clue provided by Newton that the laws of nature can be expressed in the form of differential equations.

More Examples of Clues in Scientific Inquiry

Consider another example from science: the principle of inertia (one of the most important principles in physics). Galileo is generally credited with this discovery. However, some scholars dispute whether or not Galileo reached a fully modern understanding of inertia, but the point here is not to settle that question. The point here is to illustrate how clues function in reaching scientific insights.

The principle of inertia states that an object in motion will stay in motion unless acted on by an external force and that an object at rest will remain at rest unless acted on by an external force. This notion runs contrary to everyday experience. For instance, when I kick a ball across a flat surface such as a soccer field, it gradually slows down and comes to a stop unless I kick it again. Based on observations like this, Aristotle believed that things in motion naturally tend toward a state of rest. If everyday

---

[35] Morris Kline, *Mathematics and the Physical World*, "So many physical principles are most effectively formulated as differential equations that nature and God are often credited with speaking in terms of them," pages 422–423.
[36] Partial differential equations are differential equations with more than one independent variable.
[37] Albert Einstein, *Ideas and Opinions*, page 268.

experience confirms this belief, how did Galileo reach the opposite conclusion: that in the absence of resistance, an object in motion never stops?

Galileo's groundbreaking insight began with an observation. He observed that decreasing the resistance encountered by moving objects increases the period of time that such objects remain in motion. For instance, a marble rolls across a glass surface for a longer period of time than across a rougher surface such as carpet, because glass resists the motion of a rolling marble less than carpet. Consider a question: if the length of time that an object remains in motion can be increased by decreasing the resistance it encounters, how long would an object remain in motion if the resistance disappeared entirely? Galileo reasoned that such motion would be eternal:

> "Thus a ship ... having once received some impetus through the tranquil sea, would move continually around our globe without ever stopping...if...all extrinsic impediments could be removed"[38]

An Experimental Supposition Can Be a Clue

Galileo performed what is now called a "thought experiment," which consists of following up on the logical implications of an experimental supposition. Galileo supposed a world without resistance and reasoned that, in such a world, an object, once moved, would not stop. Galileo's experimental supposition, what motion would be like without resistance, was

---

[38] Stillman Drake, *Galileo Studies*, page 251.

his clue, his thread in the labyrinth of understanding inertia. By following up on that clue, he extended what he knew to include what he did not know—he extended his knowledge that decreasing resistance increases the length of time an object remains in motion to include the knowledge that when resistance is absent, movement is eternal.

Physical Observations Can Be Clues

Consider another example of how clues guide scientists toward discovery—the principle of relativity. Einstein did not discover the core notion of relativity, Galileo did. At the time Galileo was living, many people believed that the sun revolved around the earth. Galileo, on the other hand, believed with Copernicus that the earth revolved around the sun. Those who believed that the earth did not move made the following argument:

> Anyone can see that a stone dropped from a high tower falls to the base of the tower. Now, this would not happen if the earth were moving. Why is that? If the earth were moving, then the base of the tower, which rests on the earth, would be moving too. But if the base of the tower were moving as the stone was falling, then by the time the stone hit the ground, the base of the tower should have moved out of the way. Therefore, if the earth is moving, the stone should not fall to the base of the tower. And since stones dropped from a high tower *do* fall to the base, the earth cannot be moving.

Galileo set out to prove this argument wrong and to this effect he made a striking observation about falling objects. To illustrate Galileo's brilliant observation, consider a ship at sea that is moving at a constant speed and *notice how objects fall to the floor of the ship exactly the same way they would fall if the ship were not moving.* The same phenomena can be observed in an airplane; if you are flying in an airplane that is not speeding up or slowing down, objects fall to the floor of the plane the same way they would fall if the plane were not moving. These examples all illustrate Galileo's principle of relativity. The principle of relativity can be stated in the following way: If you are in an enclosed space such as the hull of a ship that is moving at a constant speed and you cannot see outside, no experiment (such as dropping objects to the floor) will tell that you whether or not the ship is moving. Objects moving at the same rate as the ship will fall to the floor of the ship in exactly the same as if the ship were perfectly still. This principle is known as "Galilean relativity"; it is built into Newtonian mechanics and is sometimes called Newtonian relativity.

Based on the principle of relativity, Galileo reasoned that if the earth were moving at a constant speed, objects moving at the same rate as the earth would fall to the ground exactly the same way they would fall as if the earth were perfectly still. Therefore, the fact that stones dropped from a high tower fall to the base is not a proof that the earth does not move.

Galileo reached the notion of relativity by following up on the logical implications of a clue - his observation that objects moving together at a constant speed behave no differently from when they are rest.

Clues that Guided Einstein to the Discovery

Let's consider one more particular example from science: the Special Theory of Relativity. In making this monumental discovery, Einstein was guided by clues from the past. One such clue was the principle of Galilean relativity. Even before Einstein formulated the Special Theory of Relativity, he was convinced that the basic notion of relativity—that there is no way to determine whether or not you are moving in an enclosed space—was essentially correct.[39]

The core idea of relativity originally conceived by Galileo and built into Newtonian mechanics was Einstein's clue,[40] his thread in the labyrinth of understanding Special Relativity. This was not Einstein's only clue from the past—the discovery that light moves at a constant speed along with the anticipation that a scientific understanding of motion would be expressed in the form of differential equations were also clues that guided Einstein to discovery.

Einstein's discovery of Special Relativity also involved a dramatic revision of the traditional ideas of time and space. But the point of the present section is to illustrate how clues function in scientific discovery. A more careful consideration of how the notions of time and space were changed by Relativity will be

---

[39] Jeremy Bernstein, *Albert Einstein and the Frontiers of Physics*, page 54.
[40] Stephen Hawking, *A Brief History of Time*, "The fundamental postulate of relativity, as it was called, was that the laws of science should be the same for all freely moving observers, no matter what their speed. This was true Newton's laws of motion, but now the idea was extended to include Maxwell's theory and the speed of light: all observers should measure the same speed for light, no matter how fast they are moving," pages 20–21.

treated in Chapter 7 where Einstein's theory is compared and contrasted with Newton's.

Kuhn Neglected the Importance of Clues

We have been discussing the nature of clues, which are the connection or the link between the known and the unknown. Since clues connect the known to the unknown, the extent that discoveries are guided by clues is the extent that new scientific knowledge is connected to past knowledge. Moreover, since Kuhn's paradigm shift is the case where new knowledge is not connected to the past, the extent that clues lead to new discoveries is the extent to which paradigm shifts do *not* occur.

Kuhn excluded the important function of clues from his whole account of scientific discovery, making the investigator appear literally clueless. By virtue of this omission, Kuhn made the nature of scientific discovery appear to be a matter of chance. He is not alone in viewing discovery this way. According to Bernard Cohen and George E. Smith, it is a misconception of discovery known as, "The bright idea and everything-fell-into-place-myth."

> "On this misconception, the key to successful science is for someone to come along who almost magically devises a new way of thinking about some relevant aspect of the world and who is then somehow able to see almost immediately how effective this new way of thinking is going to prove in the long run."[41]

---

[41] William H. Cropper, *Great Physicists*, page 18.

"The bright idea and everything-fell-into-place-myth" is often associated with Newton, who, according to a popular story, came up with his idea of gravity by the sight of a falling apple. But this account unjustly fails to recognize the brilliant reasoning and sheer intellectual stamina involved in Newton's discovery of universal gravitation. When asked how he had discovered his theory of universal gravitation, his answer was "by thinking on it continuously."[42] Newton did not have a magical ability to make scientific discoveries:

> "Newton was exceptional not because he had a capacity to leap to correct answers, but because of the speed and tenacity with which he would proceed step-by-step through a train of inquiry, putting questions to himself, working out answers to these questions, and then raising further questions through reflecting on these answers."[43]

Newton provides a good example of how making use of clues and scientific reasoning can lead to new discoveries. His procedure was to ask questions guided by clues based on physical observation and then follow up on the logical implications of those clues[44] and thereby arrive at a new understanding that would serve as the starting point for more questions.

---

[42] Bernard Cohen and George E. Smith, *The Cambridge Companion to Newton*, page 8.
[43] Bernard Cohen and George E. Smith, *The Cambridge Companion to Newton*, page 8.
[44] Isaac Newton, *The Principia*, "In this philosophy particular propositions are inferred from the phenomena, and afterwards rendered general by induction. Thus it was that the impenetrability, the mobility, and the impulsive force of bodies and the laws of motion and gravitation were discovered..." page 443.

The Importance of Clues to Scientific Inquiry

With regard to discovery, science is not without its own brand of drama. A new groundbreaking scientific insight into the physical world is the discovery of a relation between things apparently un-related and, consequently, can be quite surprising and sometimes even melodramatic. And yet, is it really plausible to suppose that the genius of scientific discovery is simply a matter of chance? Is it not more reasonable to suppose that discovering a connection between things apparently unconnected is more likely to be achieved through an intelligent use of clues?

A scientific inquiry, like a criminal investigation, cannot proceed if there is nowhere to begin. But possessing a clue is not enough; one must follow up on a clue. For Theseus, following up on a clue meant following the thread back out of the labyrinth. For a scientist, following up on a clue means following a thread of logic. Certainly, this was the case with the discoveries of Galileo, Newton and Einstein that we have examined. Although we have only offered a few instances of how clues lead to discovery, perhaps this has been sufficient to establish the important role that clues play in scientific investigations.

Kuhn omitted the role that clues play in scientific discovery with a twofold effect. First, he makes the process of discovery itself seem random and to that extent irrational. Second, he also fails to indicate the sense in which new discoveries can be connected to the past.

Summary

(1) Scientific inquiry is most effective when the desire to understand is not unfavorably influenced by other desire.
(2) A desire to know is useless without a capacity for insight.
(3) Intelligent inquiry involves the use of clues.
(4) Making use of clues requires the ability to reason.
(5) Clues that lead to discovery extend what is known.
(6) The clue of mathematics is essential to modern science.

Chapter 4

Rational Consciousness

In an effort to dispel Kuhn's irrational depiction of science, we have been illustrating how scientific activity follows the Pattern of Rational Intelligence:

(1) Empirical Consciousness (awareness of sense perceptions)
(2) Intelligent Consciousness (asking questions and forming an understanding)
(3) Rational Consciousness (determining correctness by weighing evidence)

Just as the ball of thread aided Theseus in making his way through the labyrinth, clues help scientists to make their way through the labyrinth of understanding the universe. However, the investigation is not complete once a theoretical understanding has been reached. Many theories can offer to explain the same phenomena, but which one is correct?

The Nature of Judgments

Correctness is decided by *judging* that the proposed understanding is supported by sufficient evidence. A scientific investigation, like a criminal investigation, culminates in judgment. Judging involves weighing evidence; judging answers the question, is it so?[45] Detectives must raise this question in the

---

[45] Lonergan distinguishes between two basic kinds of questions: questions for intelligence and questions for reflection. The former seeks understanding, and

course of a criminal investigation. Once an understanding has been reached of who committed a crime, they must ask if there is sufficient evidence to convict the suspect. If there is sufficient evidence, then one must answer yes, and if not, one must answer no. It is always the case of answering yes or no to the question, "Is it so?"

Probability Judgments

Let's consider the nature of judgment in greater detail by exploring another type of judgment: a probability judgment.[46] Suppose someone asks me, "Is your brother home?" I may not know for certain, and thus I can reply, "I do not know." But suppose they ask is your bother *probably* home? Do I have sufficient evidence for *that* claim? I must weigh facts. Suppose I just got off the phone with my brother, and he said he was home. He might have lied, or I might be delusional. But I have never known my brother to lie or myself to be delusional. Furthermore, I happen to know that my brother spends most of his time at home. So, when confronted with the question, "Is your brother *probably* home?" I must answer yes.

Good Judgments

Judging is an activity distinct from understanding. A psychologist may possess a fine *theoretical* understanding of psychology, but determining the precise nature of a patient's *actual* condition requires good judgment, which is the stamp of

---

the latter seeks a yes or no to the question: Is it so? Bernard Lonergan, *Insight*, page 297.
[46] Bernard Lonergan, *Insight*, page 324.

high quality intellectual work. According to Lonergan the following factors contribute toward making good judgments:

(1) A Desire to Understand Undistorted by Other Desire.
(2) Intelligence.
(3) Experience.

A Desire to Understand Undistorted by Other Desire

Judgments can be distorted by desires other than the desire to understand correctly. For example, a judgment can be rash. We can leap to conclusions because we *want* the conclusion to be true regardless of facts. We can also want the conclusion to be false, and this can make people hesitate to draw a conclusion even in the face of overwhelming evidence. Our ability to judge is better when we are not unduly influenced by our personal likes and dislikes.

Intelligence

Being a good judge requires intelligence. What does it mean to be intelligent? Recall that Lonergan defines human intelligence in terms of a capacity for insight; the absence of insight impairs the ability to judge.

Experience

Good judgment requires experience. Experienced judgments are more likely to be correct than inexperienced judgments. A child may be very intelligent but may lack the necessary experience to make important decisions. When seeking medical advice, we don't ask just anyone; we ask doctors, not because they are infallible, but because doctors have a greater likelihood of making sound medical judgments due to their medical experience. Nonetheless, we also judge doctors to see if there is anything about them to make us doubt the soundness of their medical judgments. Have they made serious mistakes in the past? Are they reliable? Answers to these questions help us determine the soundness of their medical judgments.

The Basis for Judging a Scientific Theory to Be Correct

The decision to adopt a scientific theory requires good scientific judgment. Still, no matter how many experiments a scientist performs, no matter how much data is collected, it is impossible to gather all the pertinent facts—one cannot study every atom in the universe nor investigate every ray of light. For this reason, scientific judgments are, as a rule, only probability judgments.

To illustrate what's involved in proving a scientific theory to be correct, let us return to Galileo's discovery that a free fall is a constant acceleration. A simple and yet profound difference between Aristotle and Galileo is this: While Aristotle sought to understand the nature of a free fall by the *direct* testimony of his senses, Galileo *measured* what he observed and then sought to understand how the measurements relate mathematically to one another. Galileo thereby arrived at a modern scientific explanation, which is a mathematical law or rule that relates measurements. Verifying such a formula is the basis for judging a modern theory to be correct. Hence, the function of measurement is absolutely foundational to modern science.

The High Significance of Measurement

Why is measurement essential to science? It is because the activity of relating measurements has resulted in (possibly) the most powerful and reliable knowledge the world has ever known in the form of empirical equations. All modern technology, including telephones, traffic lights, rockets, computers, radios, television, electric appliances, cars, and jet airlines, are made possible through the function of measurement in the sciences.

Kuhn did not discuss the importance of measurement in his book *The Structure of Scientific Revolutions*, and his later remarks on the subject are essentially dismissive. His thoughts on this subject can be found in a book called *The Essential Tension*, in a chapter entitled "The Function of Measurement in Modern Physical Science." After admitting the importance of quantitative methods to modern science, Kuhn made the following statement about the function of measurement:

> "...our most prevalent notions both about the function of measurement and its special efficacy are derived largely from myth"

The same chapter concludes with this statement:

> "...can we conclude anything at all? I venture the following paradox: The full and intimate quantification of any science is a consummation to be devoutly wished. Nevertheless, it is not a consummation that can be effectively sought by measuring"

Kuhn acknowledged that his opinion about measurement is paradoxical without bothering to explain the paradox—he is denying the importance of measuring while at the same time insisting on the importance of mathematics. But the only significance of mathematics to modern science is that it quantifies relationships between measurements. If one dismisses the function of measuring, then mathematics is **absolutely useless** to the sciences.

By disregarding the significance of measurement, Kuhn failed to distinguish pre-scientific thought from modern science. For instance, he claimed that a paradigm shift occurred between Aristotle and Galileo. However, the transition from Aristotle to Galileo was not a paradigm shift—not the transition from one equally valid scientific theory to another. It was the transition from philosophy to modern science. It is true that there are similarities between Galileo and Aristotle. Both inquired, observed and reasoned about the nature of a free fall. But Galileo took additional steps. He:

(1) Measured.
(2) Experimented.
(3) Sought a mathematical correlation between measurements.

It is *specifically* these additional steps that make modern technology possible. Consequently, it is these three steps that sharply distinguish a modern inquiry from a pre-scientific one. Kuhn did not make this clear. Therefore, it was misleading of Kuhn to treat Aristotle as though he were a scientist on par with

Galileo. Aristotle wasn't a scientist in the modern sense of the word; he was a bridge between philosophy and science.

Kuhn also claimed that the transition from the phlogiston theory to Lavoisier's theory of combustion was a paradigm shift. What was the Phlogiston theory? It was a theory of combustion— an explanation for what happens when things burn. Generally, when something is burned, it weighs less than before it was burned. This loss of weight was explained by the release of a hypothetical substance called phlogiston. The phlogiston advocates believed that wood weighs less after you burn it because of the phlogiston released into the air.

One major problem with the phlogiston theory was that certain metals actually gain weight after being burned. To explain this, the advocates of the phlogiston theory suggested that in some metals phlogiston had negative weight. Hence, phlogiston could have positive weight, negative weight and in some cases even zero weight.

Prior to Lavoisier (1743–1794), most chemists did not recognize the significance of measurement. It was through measuring and careful quantitative analysis that Lavoisier was finally able to account for all of the products and reactants in combustion, putting a decisive end to the phlogiston doctrine. Phlogiston theory was pre-scientific insofar as the great importance of measuring was not generally appreciated. The shift from the phlogiston theory to the theory of combustion was not a paradigm shift; it was not the transition from one equally valid scientific theory to another. It was the shift from a pre-scientific understanding of combustion to a modern scientific explanation—one that establishes relations between measurements.[47]

It is an oversimplification to say that the theory of phlogiston was categorically wrong. The advocates of phlogiston theory were vaguely groping toward an understanding of combustion. They were not wrong in supposing that some substance was being added or subtracted from combustible materials. Nor were they far from Lavoisier's discovery. In fact, if one defines phlogiston as negative oxygen, the concept of phlogiston satisfies the modern requirements for a scientifically valid theory of combustion. But, there was another side to the phlogiston problem. Phlogiston was frequently a vague concept. For some, it was a substance, for others, it was closer to what we mean by "energy." Lavoisier's discovery answered questions about the composition and weight involved in combustion but did not address the energy relations involved in the process of combustion.[48] In this sense, phlogiston theory sometimes addressed an aspect of combustion that Lavoisier did not.

By establishing relationships between measurements, Galileo discovered that a free fall is a constant acceleration and became known as the Father of Modern Science. By establishing relationships between measurements, Lavoisier discovered the role of oxygen in combustion and became known as the Father of Modern Chemistry. It was through the function of measurement

---

[47] John Gribbon, *The Scientists*, "The phlogiston model worked, after a fashion, as long as chemistry was a vague, qualitative science. But as soon as Black and his successors started making accurate measurements of what was going on, the phlogiston doctrine was doomed," page 258.

[48] Douglas Allchin, *Phlogiston after Oxygen*, 1992, Ambix 39(3): 110–16.

that John Dalton produced the first table of atomic weights and became known as the Father of Modern Atomic Theory:

> "It is a striking fact that the history of each of the sciences shows continuity back to its first use of measurement, before which it exhibits no ancestry but metaphysics. That explains why Galileo's science was stoutly opposed by nearly every philosopher of his time, he having made it as nearly free from metaphysics as he could. That was achieved by measurements."[49]

Verifying a formula that relates measurements is the basis for determining if a modern theory is correct. Arguing against this point, Kuhn claimed that accuracy alone is *not* always decisive in judging a theory to be correct.[50] To illustrate his point, he argued that the choice between Ptolemy's earth-centered astronomy and Copernican sun-centered astronomy cannot be made based solely on accuracy. Kuhn was certainly right about this. If you lived around the time of Copernicus, judging which astronomical theory was correct would surely be difficult since the Copernican sun-centered model was not quantitatively more precise that the Ptolemaic earth-centered model. Nevertheless, sixty years later, all the pertinent astronomical questions that centered around Ptolemy and Copernicus were settled by Tyco Brahe spending long years making precise measurements and Kepler's mathematical insight into those measurements, revealing that the planets move not in circular but elliptical paths around the sun. Quantum mechanics is perhaps the clearest example of how

---

[49] William H. Cropper, *Great Physicists*, page 17.
[50] Thomas Kuhn, *The Essential Tension*, page 322.

verifying relations between measurements is *the* basis by which a modern theory is judged to be correct. Despite the perpetual lack of agreement about which theoretical model of quantum theory is correct, quantum theory was accepted as valid because of its outstanding success at predicting the outcomes of measurements.

Consider another modern example of how measurement is decisive in determining the correctness of scientific theory: the perihelion of Mercury. What does *perihelion* mean? The perihelion is the point nearest the sun in the orbit of a planet or comet. Every time Mercury orbits the sun, the point where it passes closest to the sun changes by a very small degree. The Newtonian calculation of the change of the perihelion of Mercury was very accurate, but a little off. To be precise, it was off by forty-three seconds of arc per century. Einstein's general theory of relativity corrects this error. The decision to adopt the general theory of relativity was intrinsically rational; it was more precise than Newton's theory.

Summary

Experimental verification of relations between measurements is the basis for judging a modern scientific theory to be correct. Without the use of measurement, modern technology would not be possible and science would not significantly differ from philosophy. A strong theme in Kuhn's writing is the proposition that there are no impartial or objective criteria for determining whether or not a scientific theory is correct. Certainly without measurement, the decision becomes much more subjective. Measurement is the most objective component of theory choice. It is precisely because Kuhn dismissed the importance of measurement that he cannot clearly distinguish pre-scientific thought from modern science. Most of the examples that he provides, which were intended to illustrate the subjective nature of theory choice, are simply cases where the ability to measure was lacking or the importance of measuring unappreciated.

Chapter 5

The Nature of Scientific Discovery

Let us now summarize the rational view of scientific discovery that has been articulated in the last section of this work.

Wonder—the Source of all Science

We usually acquire knowledge because we want it. Sometimes we want to know for practical reasons, and other times we are simply curious. Such curiosity can occasionally be a passionate interest. It is this kind of desire that commonly inspires great scientific discoveries—an observation that has been made by some of the greatest minds in human history. Aristotle, for instance, proclaimed wonder to be the source of all science and philosophy, and Einstein believed that without a passion for comprehension science would be impossible.

The Objective of Science

Science begins with a desire to know, but not a desire to know just anything. The objective of modern scientific inquiry is restricted to the data of sensory experience; consequently, the process of scientific discovery begins by seeking insight into what can be observed and measured and concludes with observations and measurements. The high importance of physical observation was not always recognized. It was precisely because Galileo recognized the monumental significance of observation that Einstein considered him to be the Father of Modern Science.

The Scientific Method

The scientific method is essentially the same pattern of rationality found in everyday life that Lonergan identified as observing, understanding and judging. According to Lonergan, this pattern involves three distinct levels of consciousness: Empirical, Intelligent and Rational. The desire to know sets the whole process in motion and moves us from one level to the next. Desiring to understand the *empirical* world of sensory experience inspires the *intelligent* activity of asking questions and forming understanding; desiring to know if such understanding is correct inspires the *rational* activity of weighing evidence in an effort to answer the question: Is it so?

Einstein once called science a refinement of everyday thinking. He was right because, as Lonergan observed, the pattern everyday thinking essentially consists of *observing*, *understanding* and *judging*. How is science a refinement of this pattern? One can almost answer in a single word—measurement. It is through the function of measurement that scientific thought attains precision and exactness. Scientific *observation* is directed toward measuring, which is a numerically precise type of observation. Scientific *understanding* is directed toward obtaining a mathematical insight into measurements, and such understanding is *judged* to be correct only if the relations between measurements can be precisely verified. The chart below shows how scientific thinking is simply an expansion of the same pattern of rationality present in everyday common sense:

Consciousness	Common Sense	Scientific Thought
Empirical	Observing	Measuring
Intelligent	Seeking insights into observations	Seeking mathematical Insights into measurements
Rational	Insights are judged to be correct based on sufficient reason	Insights are judged to be correct based on verifying relations between measurements

The Rational Nature of Discovery

How are new discoveries made? There is no full and complete answer to this question. Therefore, in an important sense, the nature of discovery remains mysterious. Nonetheless, despite the mystery involved in making new discoveries, Lonergan observed that the following conditions increase the likelihood of achieving insight:

(1) The Desire to Know.
(2) The Capacity for Insight.
(3) Clues.
(4) The Ability to Reason.

The Desire to Know

Wanting to comprehend a subject that interests us increases the likelihood that we will pursue a comprehension of it. But other desires can interfere with our pursuit of understanding. For instance, understanding can be *unwanted* because it is too painful to think about, as is sometimes the case when an alcoholic refuses to admit the existence of a drinking problem. Hence the chances of reaching insight are further improved when personal wants, fears and needs do not unfavorably influence the desire to know.

The Capacity for Insight

Still, even a desire for knowledge that is entirely free from personal bias would be useless without intelligence. According to Lonergan, intelligence essentially resides in the capacity for insight, of which there can be many kinds. There are psychological insights, poetic insights, medical insights, philosophical insights, literary insights, practical insights, and so on. The ultimate goal of modern scientific inquiry is to obtain a mathematical insight into the measurable aspects of the world apprehended through sensory experience.

Clues

Still, even being highly intelligent is no guarantee that a search for understanding will succeed. So how does one get from inquiry to insight? In everyday life, people make use of clues. Whether you are a doctor diagnosing an illness, a mechanic trouble-shooting an inoperative car, or a detective trying to find the culprit of a crime, finding the right clues are an essential part of an intelligent inquiry. Scientific inquiry is no exception. Without clues, the possibility of achieving insight is reduced to a matter of chance. While such arbitrariness sometimes plays a part in the process of discovery, the scientific method could not consistently yield insights into the physical world without an intelligent use of clues.

The Ability to Reason

Still, possessing a clue would be useless without the ability to reason. It is true that doctors, mechanics and detectives need clues in order to make discoveries, but they also need the ability to draw out the logical implications of clues. In this respect scientific investigations are no different from criminal investigations, medical inquiries or automotive trouble shooting.

The Importance of Technology

There remains one more important aspect of scientific discovery to be discussed: technology.[51] In the first place, technology serves as a principle of verification. Building machines based on equations serves to verify the validity of existing theory or reveal cases where nature does not behave as theory predicts, thereby providing new areas to investigate. In effect, applied science constantly puts theoretical expectations to the test without necessarily intending to do so.

Through technology we also become more aware of the universe. For instance, Aristotle thought that the moon was a perfect sphere. Imagine Galileo's surprise when he looked through a telescope and saw mountains on the moon. Thus, through the invention of the telescope, we became aware of mountains on the moon just as the invention of the microscope made us aware of microorganisms. Only a small part of the universe falls within the range of human sensory experience. Technology enables us to extend our natural powers of

---

[51] For an in-depth discussion on the role of technology in scientific thought see Bernard Lonergan, *Insight*, page 97.

observation far beyond the limits of our biological sensing mechanisms, and in this respect constitutes a very real expansion of human consciousness. Moreover, technology allows us to make precise measurements, which provides more accurate data, and more opportunities for insight, which can lead to an improved technology and a greater ability to measure. Thus, scientific progress partly depends upon the technology that it creates.

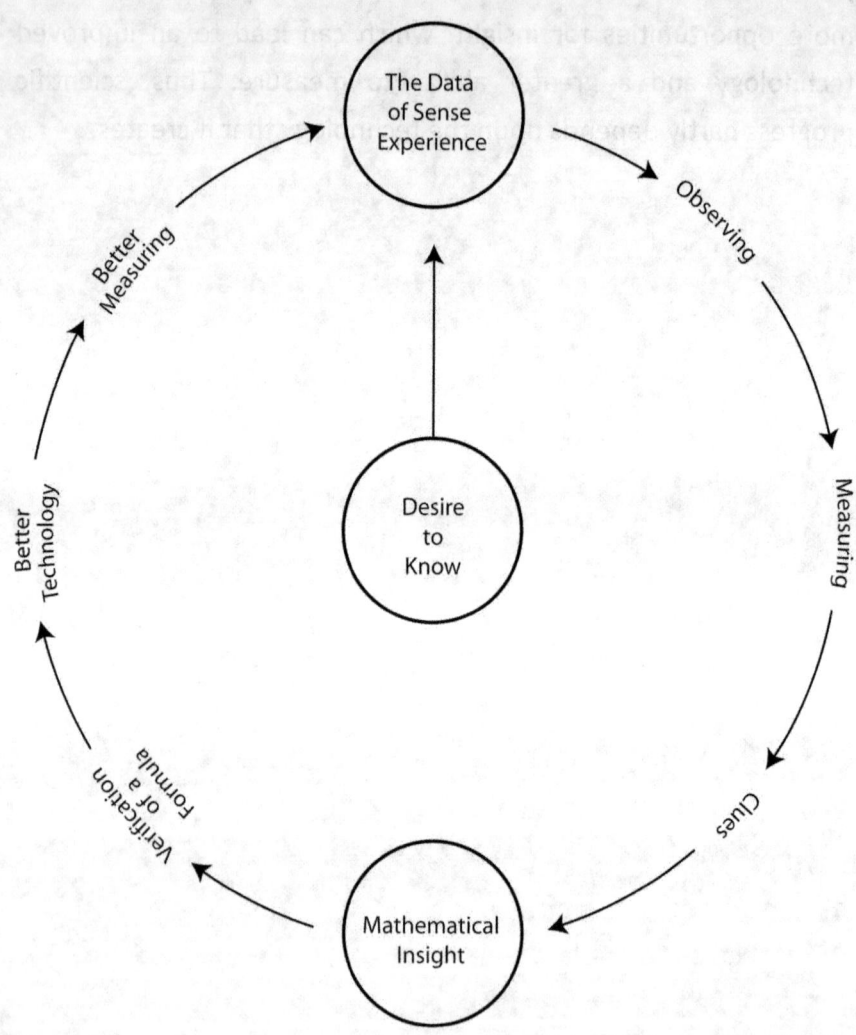

Distorting the Rational Nature of Discovery

With regard to scientific discovery, we paid special attention to the function of clues. Not only do clues reduce an element of chance in the process of discovery, they cast light on the rational nature of intelligent inquiry by illustrating how new discoveries can be reached by drawing out the logical implications of clues. Thus, we argued that inspired by a desire to know, guided by clues and the ability to reason, scientists make their way through the labyrinth of understanding nature. Compare this pattern of discovery with Kuhn's position that the emergence of a new theory cannot occur without a *crisis*, which he describes as a "professional insecurity" that arises from the perception of nature not behaving as expected. While feelings of insecurity may sometimes play a role in scientific discovery, does it really make sense to suppose that such fear is necessary?

When Kuhn described a scientist searching for a new theory, he heavily emphasized the elements of mystery[52] and chance.[53] Moreover, Kuhn included no discussion of the function of clues in scientific investigations, which distorted a rational comprehension of science in two ways. First, by excluding the function of clues from the process of discovery, Kuhn obscured the sense in which new ideas are connected to the past. Secondly,

---

[52] "What the nature of that final state is—how an individual invents (or finds he has invented) a new way of giving order to data now all assembled—must remain inscrutable and may be permanently so,'" Thomas Kuhn, *The Structure of Scientific Revolutions*, page 91.

[53] Kuhn describes the search for a new theory: "He will, in the first place, often seem a man searching at random, trying experiments just to see what will happen, looking for an effect whose nature he cannot quite guess," Thomas Kuhn, *The Structure of Scientific Revolutions*, page 87.

he makes the scientific investigator appear literally clueless and to that extent irrational.

There are other ways that Kuhn distorted the nature of science. For example, Kuhn claimed that the goal of science periodically changes[54] and yet failed to clarify that the unchanging object of scientific investigation is the data of sensory experience. Kuhn also claimed that the scientific method periodically changes, but failed to clarify that the unchanging foundation of the scientific method is the same pattern of rational activity to be found in everyday life, which consists of observing, understanding and judging. Modern science is a refinement this pattern through the function of measurement, which may be described in the following manner: measuring is a refinement of observation; mathematical insight into the measurements is a refinement of understanding; and verifying relations between measurements is a refinement of judgment. But Kuhn dismissed the significance of measurement, thereby obscuring the distinction between pre-scientific thinking and modern science. However, the significance of his dismissal of measurement cannot be fully appreciated until we address the relation of measurement to the nature of scientific knowledge, which will be discussed in the next chapter.

---

[54] Ibid., page 85.

Knowledge

Chapter 6

The Nature of Scientific Knowledge

Timeless Equations

The timeless character of scientific knowledge resides in the empirical equations of modern science, which are mathematical formulas that represent relationships between measurements—a relation that must be repeatable before it may be accepted as valid. Hence, the requirement for constancy is built into the scientific method itself. Once such an equation has been tested and verified, it *always* functions the same way even though there may eventually be a different explanation for why the equation works. For instance, the empirical equations of Newtonian physics are still used today even though the modern explanation for why they work differs from Newton's explanation.[55] It is precisely because empirical equations are constant that reliable machines can be built based on them. Even those who reject the very possibility of scientific objectivity still use machines based on scientific equations with notable confidence.

How Things Relate Numerically to One Another

The simple act of measuring was perhaps the most profound difference between Aristotle and Galileo, for Aristotle sought an insight into the nature of a free fall based on what he observed, whereas Galileo *measured* what he observed and then

---

[55] Newton correctly stated that a change in momentum is proportional to the external force but incorrectly assumed that time, space and mass are invariant.

sought an insight into the measurements. This is the key to understanding modern science for, according to Lonergan, it is through the act of measuring that:

> "We move away from colors as seen, from sounds as heard, from heat and pressure as felt. In their place we determine the numbers named measurements. In virtue of this substitution we are able to turn from the relations of sensible terms, which are correlative to our senses, to the relations of numbers, which are correlative to one another. Such is the fundamental significance and function of measurement."[56]

Through the function of measurement, modern science acquires knowledge of how things relate numerically to one another. This knowledge takes the form of empirical equations that always function the same way and upon which all modern technology depends. Who could have guessed that the simple act of measuring would be instrumental to unlocking the secrets of the universe?

A Closer and Closer Approximation of the Truth

Modern science has a history of being able to predict the behavior of nature with greater and greater accuracy. For instance, a brief snapshot of the development of astronomical knowledge would look like this: Tyco Brahe spent long years diligently observing and recording the positions of the planets,

---

[56] Bernard Lonergan, *Insight*, page 188–189.

providing Kepler with an opportunity to seek a mathematical insight into Brahe's measurements and subsequently prove that the shape of a planetary orbit is not a circle but an ellipse; Newton discovered equations that could accurately predict the position of a planet at a given instant, and Einstein's theory of relativity yielded an even more precise knowledge of planetary motion. The empirical equations of modern science predict with increasing accuracy an increasing range of phenomena. In this mathematical sense, the history of modern science has been a closer and closer approximation of the truth.

Scientific Knowledge Is Rational

The rational character of modern scientific knowledge can be most clearly seen in that it assumes a mathematical form, which is very useful from a scientific standpoint. Once a few key equations have been discovered that establish a mathematical relation between the measurable aspects of natural phenomena, it is possible to employ deductive mathematical reasoning and thereby acquire an enormous amount of information about such phenomena. This provides powerful evidence that the universe is a logical mathematical structure that, to some extent, is within the reach of rational inquiry. Moreover, once a system of scientific laws have been logically deduced, engineers and applied scientists must again employ deductive reasoning in order to determine which laws are to be used in a given concrete situation.

The Search for New Knowledge

With regard to the search for new knowledge, it is worth noting that scientists who make discoveries that lead to useful technological developments are consistently guided by two basic anticipations:

(1) Scientists anticipate a mathematical relation to exist between measurements.
(2) Scientists anticipate that their discoveries will be *universal* in some respect.[57] (For instance, the limitations of measurement expressed by the Heisenberg Uncertainty Principle are valid for all electrons.)

Distorting the Meaning of Modern Science

Kuhn argued that scientific knowledge fundamentally changes; thus, what was true yesterday may be false today and what is true today may be false tomorrow. The very expression "scientific revolution" in the title of his book *The Structure of Scientific Revolutions* implies that scientific knowledge can be overturned by means of a revolution in the same way that a government can be overturned. Kuhn recognized that this analogy between politics and science required justification;[58] therefore, he argued that fundamental changes in political systems are akin to fundamental changes in scientific systems. However, one essential difference that Kuhn excluded from his discussion is this: when one political system replaces another, the old laws *cease* to function; whereas, when a modern scientific system replaces another, the old mathematical laws *continue* to function.

---

[57] Bernard Lonergan, *Insight*, page 61.
[58] "Why should a change of paradigm be called a revolution? In the face of the vast and essential differences between political and scientific thought what parallelism can justify the metaphor that finds revolutions in both?" Thomas Kuhn, *The Structure of Scientific Revolutions*, page 92.

Generally speaking, Kuhn paid exclusive attention to the areas of science that exhibit discontinuity and remained silent about areas that exhibit continuity and connectedness. Thus, with regard to discovery, he emphasized the element of chance[59] and neglected the important function of clues. With regard to knowledge, he discussed the aspects of scientific theory that are subject to revision, but not the empirical equations that are not. But Kuhn's single greatest distortion of science was his dismissal of the importance of measurement, which plays a central role in the all the essential moments of modern science, including observation, understanding, judgment and the empirical equations that result from these activities.

By discounting the importance of measurement, Kuhn blurred the distinction between modern science and pre-scientific thinking. He thereby failed to mark the beginning of modern science with Galileo, the beginning of modern chemistry with Lavoisier, and the beginning of modern atomic theory with John Dalton. He mistakenly presented the change from Aristotle to Galileo as paradigm shift rather than a transition from philosophy to science. He also erroneously treats the change from phlogiston theory to Lavoisier's theory of combustion as a paradigm shift rather than the transition from a pre-scientific description of nature to a modern scientific explanation—one that establishes mathematical relations between the measurements.

Moreover, verifying a formula that relates measurement supplies the most objective component in determining the probable truth of a scientific theory. If one excludes the function

---

[59] Thomas Kuhn, *The Structure of Scientific Revolutions*, page 87.

of measurement, as Kuhn has done, then the rationale for adopting a new theory is indeed more subjective. Furthermore, neglecting the significance of measurement makes it difficult if not impossible to see the rational, unchanging and cumulative nature of modern science, which is rooted in measurement. If one uproots measurement from science then the whole meaning of modern science disappears into the fog of pre-scientific thought.

Chapter 7

Einstein Builds on Newton's Theory

Modern scientific equations, once tested and verified, always function the same way, although the theoretical concepts that explain why such equations work may eventually differ from the original explanation. For example, Einstein's theoretical explanation for why Newton's equations always work differs from the original Newtonian explanation. In this chapter we will show how Kuhn exaggerated the theoretical differences between Newton and Einstein. Just as he presented discovery as motivated by anxiety rather a desire to know, just as he made new insights appear to be the outcome of chance by ignoring the use of clues, just as he obscured the meaning of modern science by neglecting the function of measurement, so he also presented Einstein's theoretical concepts as incompatible with those of Newton by ignoring the common ground between them. This common ground is important because, for Kuhn, the theoretical difference between Newton and Einstein is an exceptionally clear example of a conceptual paradigm shift:

> "The transition from Newtonian to Einsteinian mechanics illustrates with particular clarity the scientific revolution as a displacement of the conceptual network through which scientists view the world."[60]

---

[60] Thomas Kuhn, *The Structure of Scientific Revolutions*, page 102.

Kuhn stated that the conceptual differences between Newton and Einstein were so great that it is impossible to regard Einstein's theory as correct without regarding Newton's theory as incorrect.[61] What he did not say is that Einstein's opinion squarely contradicts his own. Einstein claimed that his work did *not* essentially replace Newton's theory; rather, he repeatedly[62] insisted that he built on the work of Sir Isaac Newton:

> "Let no one suppose, however, that the mighty work of Newton can really be superseded by this or any other theory, his great and lucid ideas will retain their unique significance for all time as the foundation of our whole modern conceptual structure in the sphere of natural philosophy."[63]

Surely Einstein's opinion on the relation of his own work to Newton should carry at least as much weight as that of Thomas Kuhn's opinion. Einstein was not only a great scientist, he was also very interested in the history and philosophy of science and wrote extensively not only about the nature of science, but particularly about effect of Newton's ideas on the development of theoretical physics.[64] He was, therefore, in a position to form an intelligent

---

[61] "Einstein's theory can only be accepted with the recognition that Newton was wrong," Thomas Kuhn, *The Structure of Scientific Revolutions*, page 98.

[62] Einstein frequently acknowledged his debt to the past: "The four men who laid the foundation of physics on which I have been able to construct my theory are Galileo, Newton, Maxwell, and Lorentz," Alice Calaprice, *The Expanded Quotable Einstein*, page 240.

[63] From an article entitled "What is the Theory of Relativity?" *London Times*, November 28, 1919.

[64] From an article entitled, The Mechanics of Newton and their influence on the development of Theoretical Physics," *Die Naturwissenschaften*, Vol. 15, 1927.

opinion on the subject. Einstein's conviction was that the theory of relativity was *not* a revolution:

> "We have here no revolutionary act, but the natural continuation of a line that can be traced through centuries."[65]

Since Kuhn believed that one could not accept Einstein's theory without at the same time rejecting Newton's theory, it follows that the extent to which Newtonian and Einsteinian theories agree is the extent to which Kuhn was mistaken. Let us, then, consider the areas where Einstein and Newton would agree.

Newton's Three Laws of Motion

What is often overlooked is that Newton's original three laws of motion were not equations but general ideas about the nature of motion and as general ideas, Newton's laws of motion are absolutely correct.[66]

(1) The first law is the principle of inertia. It states that an object in motion will continue to move unless acted on by

---

[65] From an article entitled "On the Theory of Relativity—Lecture at Kings College," London 1921. Published in *Mein Weltbild*, Amsterdam: Queride Verlag 1934. Einstein's view of scientific progress is plainly developmental: "The theory of relativity is nothing but another step in the centuries-old evolution of our science, one which preserves the relationships discovered in the past, deepening their insights and adding new ones," Alice Calaprice, *The Expanded Quotable Einstein*, page 239.

[66] For a more complete analysis of how Newton's laws are still essentially correct, I refer the reader to the book *Newton Vs. Relativity*, by Jean-Michel Rocard.

an external force and that an object at rest will remain at rest unless acted on by an external force.

(2) The second law states that the change in momentum is proportional to the external force.[67]

(3) The third law is that for every action there is an equal and opposite reaction.

Newtonian Laws of Conservation

From Newton's three laws of motion one can derive the following laws of conservation, all of which are essentially true for Einstein:

(1) Conservation of linear momentum
(2) Conservation of angular momentum
(3) Conservation of energy

The Principle of Relativity

One basic concept common to Newton and Einstein is the core idea of relativity, which is this: no experiment performed in an enclosed space can determine whether or not you are moving. This notion is the essence of Newtonian relativity, which states that the laws of mechanics must be the same for all observers moving at the same velocity. The same notion also appears in one of the postulates of Special Relativity, which states that the laws of physics are the same for all inertial observers ("inertial" meaning the velocity does not change).

---

[67] This general idea can be expressed as F=MA, which is not valid for relativity but can also be expressed as F=dp/dt, which is valid for relativity.

The notion of relativity has been a recurrent theme in physics since Galileo discovered it. Einstein extended Galileo's idea to include the speed light, which is always measured to move at the same speed whether an observer is moving or not. The logical consequences of extending the principle of relativity to include the constant speed of light led to a dramatic revision of the traditional notions of time, space and mass. Let us now consider the nature of these revisions.

Mass

This is what Kuhn has to say on the subject of mass:

"Newtonian mass is conserved; Einsteinian is convertible with energy. Only at low velocities may the two be measured in the same way, and even then they must not be conceived to be the same."[68]

In the above passage, Kuhn directs the attention of the reader to the central differences between a Newtonian and relativistic concept of mass; in Newtonian theory the quantity of mass does not change, whereas in the theory of relativity, the quantity of mass changes with respect to velocity. Also, the logical implications of Special Relativity led to the discovery that mass and energy are connected[69]—a connection unknown to Newtonian physics. However, in addition to differences, there are also similarities, which Kuhn did not mention. For instance, the

---

[68] Thomas Kuhn, *The Structure of Scientific Revolutions*, page 102.
[69] From an article entitled "The Problem of Space, Ether, and the Field in Physics," *Mein Weltbild*, Amsterdam: Querido Verlag, 1934.

idea of mass as the degree to which a body resists being accelerated did not change upon the adoption of the theory of relativity just as the notion of energy as the ability to do work[70] did not change.

In the passage quoted above, Kuhn also makes a brief reference to the Newtonian conservation of mass, but did not indicate how this conservation law stands in relation to the conversation law expressed in the theory of relativity. This relation may be explained in the following manner: The discovery of the connection between mass and energy resulted in the two separate Newtonian conservation laws of mass and energy being combined into a single mass-energy conservation law.

Absolute Time and Space

Here are Newton's ideas on absolute time and absolute space as stated in his own words[71]:

> "Absolute, true, and mathematical time, of itself, and from its own nature, flows equally without relation to anything external..."

> "Absolute space, in its own nature, without relation to anything external, remains always similar and immovable..."

Note that time and space are being defined "without relation to anything external." In order to clarify Newton's

---

[70] Work in this context refers a force that moves an object over a distance.
[71] Isaac Newton, *The Principia*, page 13.

meaning, imagine a river that flows at a rate of five miles an hour in relation to a riverbank where you are standing. Now imagine the same river flowing but not in relation to you or anything else—not in relation to a riverbank nor the sky nor the earth nor the sun nor space nor anything else in the universe—nothing at all. This is impossible because we cannot imagine or observe anything moving unless it is in relation to some other thing. Thus Newton's absolute time and space are *necessarily* unobservable because they are defined "without relation to anything external." This is problematic from the standpoint of empirical science because a theoretical concept must be experimentally verified by observation before it may be considered valid. This point is especially significant since Thomas Kuhn claimed that a conceptual paradigm shift occurred between Newton and Einstein. One must bear in mind that a paradigm shift can only apply to propositions that are *equally* scientific. For instance, one could not claim that a paradigm shift occurred between a theory of art and a theory of science. Now, the different concepts of time and space proposed by Newton and Einstein were not *equally* scientific propositions since Newton's absolute time and space were, by definition, non-observable concepts, whereas the notions of time and space proposed by the theory of relativity have been repeatedly verified by observation. Therefore, the change from the absolute conception of time and space to a relative one cannot qualify as a valid conceptual paradigm shift.

This analysis by no means intends to diminish the inestimable significance of Sir Isaac Newton, who numbers among the most important scientists that the world has known. Certainly Newton needed to lay some kind of conceptual groundwork regarding the basic notions of time and space; nonetheless, his definitions were immediately subject to criticism from his contemporaries. It may be that Newton's religious beliefs influenced his thoughts on this subject, since he connected absolute time and space with the eternal omnipresent nature of God.

For Newton, all measurements of time and distance should be the same for all observers. This is what it means, in the context of a Newtonian framework, to claim that time and space are absolute. Without launching into a lengthy discussion, the point is that Newton was mistaken. He believed that time and space are separate and unchanging, whereas Einstein discovered that time and space are connected and vary with respect to velocity. This is beyond question a striking difference. Nonetheless, Einstein's own opinion was that the Special Theory of Relativity did not *replace* the Newtonian concepts of time and space, but *modified* them.[72] He was aware that the extraordinary nature of his discovery was prone to misinterpretation; therefore, he insisted that the idea of a four-dimensioned physical reality was *not* a new idea:

> "In the first place we must guard against the opinion that the four-dimensionality of reality has been newly introduced for the first time by this theory. Even in classical physics the event is localized by four numbers, three spatial coordinates and a time coordinate; the totality of

---

[72] From the revised edition of *Relativity, The Special and General Theory: A popular exposition*, translated by Robert W Lawson. London: Methuen, 1954.

physical events is thus embedded in a four-dimensional continuous manifold."[73]

Newton would agree with Einstein that motion is scientifically understood in terms of a coordinate system consisting of four numbers, three spatial coordinates and the time coordinate, although, for Einstein, time and space are two aspects of one and the same thing, whereas for Newton, they are two quite different things. But the character of "absoluteness" is common to both, for under the Newtonian theoretical model, the character of absoluteness was true of time and space separately, whereas in Special Relativity, the union of the three dimensions of space together with the dimension of time resulted in a four-dimensioned physical space. which, according to Einstein, is "just as rigid and absolute as Newton's space."[74]

Non-Euclidean Geometry

One important scientific difference between the respective theoretical models of Newton and Einstein is the use of geometry. Newton's system makes use of Euclidean geometry, while Einstein makes use of what is known as non-Euclidean geometry. Still, there is a connection between Euclidean and non-Euclidean geometry, which may be explained the following way. In Euclid's famous book, *Elements,* he states five basic postulates:

---

[73] Albert Einstein, *Relativity: The Special and General Theory: A Popular Exposition*, translated by Robert W. Lawson. London: Methuen, 1954, page 148.
[74] From an article entitled "The Problem of Space, Ether, and the Field in Physics," *Mein Weltbild*, Amsterdam: Querido Verlag, 1934.

(1) A straight line may be drawn from any one point to any other point.
(2) A finite straight line may be produced to any length in a straight line.
(3) A circle may be described with any center at any distance from that center.
(4) All right angles are equal.
(5) If a straight line meets two other lines, so as to make the two interior angles on one side of it together less than two right angles, the other straight lines will meet if produced on that side on which the angles are less than two right angles.

The 5th postulate is commonly known as the "parallel postulate,"[75] which does not hold true for non-Euclidean geometries. Still, Euclid deduced the first twenty-eight theorems of his system using only the first four postulates—a group of theorems known as Absolute Geometry. These postulates hold true for the Euclidean geometry of Newtonian physics as well as the non-Euclidean geometry of the theory of relativity. The term "non-Euclidean" is therefore an unfortunate expression because it implies that non-Euclidean geometry has nothing in common with Euclidean geometry, which is false.[76]

---

[75] This was the parallel postulate as stated by Euclid himself. Notice that he does not say the parallel lines never meet—a statement proposed by later mathematicians; rather, he states the conditions under which two lines meet. One can appreciate why he may have avoided stating that parallel lines never meet—how could this proposition be physically demonstrated?

[76] Albert Einstein, *Relativity, The Special and General Theory*, page 63.

The Pythagorean Theorem

An important connection between Newton and Einstein is the use of the Pythagorean Theorem. This theorem deserves special attention because it illustrates with striking clarity the mathematically unchanging nature of physical reality that encompasses Einstein and Newton and goes beyond them. The theorem applies to all right triangles consisting of two legs and the diagonal connecting them, and may be expressed as follows:

$$D^2 = X^2 + Y^2$$

This theorem, which is at least 4000 years old, plays an important role in *all* the sciences, and there are over four hundred different proofs for it.[77] Distances can be calculated in two, three and four dimensions using the Pythagorean Theorem, as can be seen below:

Dimensions	Distance Calculation	Theory
2	$D^2 = X^2 + Y^2$	Newtonian Physics
3	$D^2 = X^2 + Y^2 + Z^2$	Newtonian Physics
4	$D^2 = X^2 + Y^2 + Z^2 - T^2$	Special Relativity

The chief difference is this: distances in Newtonian physics can be calculated using the Pythagorean Theorem applied to a two or three dimensional space, whereas in Special Relativity, distance is calculated by an extended version of the Pythagorean

---

[77] Eli Maor, *The Pythagorean Theorem*.

Theorem[78] in which the time component squared is prefixed by a negative sign rather than a positive sign. However, summing the square of the spatial components is common to Newtonian physics and Special Relativity.

Differential Equations

With regard to the application of differential equations to natural phenomena, modern physics rests squarely on the shoulders of Newton. For Einstein, differential equations express "primary realities of physics." Kuhn omitted this relevant fact, which has been noted not only by Einstein, but also is commonly acknowledged among scientists and mathematicians.

Universal Gravitation

Kuhn did not mention that both Newton and Einstein held gravitation to be universal.

Determinism

Kuhn did not mention that both theories are deterministic, meaning they absolutely affirm the law of causality, which states that every cause is necessarily followed by an effect.

---

[78] "An expanded version of the Pythagorean Theorem also played a key role in the General Theory of Relativity," Eli Maor, *The Pythagorean Theorem*, page 195.

Inertial Systems

Einstein calls attention to another feature common to Special Relativity and Newtonian mechanics that Kuhn neglected.

> "Special relativity has this in common with Newtonian mechanics: The laws of both theories are supposed to hold only with respect to certain coordinate systems: those known as "inertial systems."[79]

The Equivalence of Inertial and Gravitational Mass

Einstein's theoretical concepts are not only connected to Newtonian theory in many ways, but also his scientific work actually confirmed certain features of Newton's thought, including what is commonly known as the equivalence of inertial and gravitational mass. This can be described the following way:

> *Gravitational mass* can be computed by using an ordinary scale that compares two objects where the weight of one object is known and the other unknown and where the force of gravity remains constant, which is usually the case on our planet.

> *Inertial mass* is determined by 1) applying a known force to an object; 2) measuring the resulting acceleration; 3) calculating the mass by using a formula that relates force, acceleration and mass.

---

[79] From an article entitled "On the Generalized Theory of Gravitation," *Scientific American*, Vol. 182, No. 4, 1950.

In Newtonian theory, inertial mass and gravitational mass are always equal—a fact that, according to Einstein, is the basis for the general theory of relativity.[80]

Also, Einstein played an important role in confirming the existence of atoms—another feature of Newtonian thought. Kuhn omitted Einstein's confirmation of Newtonian scientific concepts from his account of the subject.

A Critical Difference

The critical difference between Newton and Einstein is that the quantities of time, space and mass, once thought by Newton to be invariant, were discovered to be connected and vary with respect to velocity. And yet, it is important to recognize that once these variable quantities are included in the results of Newton's equations, Newton turns out to be correct. Thus Newton must have been *partially* right, otherwise including Einstein's relativistic correction factor could not result in a correct number. In other words, Newton was not *simply* wrong; *Newtonian mechanics still work*. There is clearly a *proximal* relationship between Newtonian scientific concepts and physical reality.

It is worth noting that Newton did not regard his theory as *absolutely* correct but only *provisionally* so, for he anticipated the possibility of uncovering data that his theory may not adequately explain:

---

[80] From an Article entitled "On the Theory of Relativity – Lecture at Kings College," London 1921. Published in *Mein Weltbild*, Amsterdam: Queride Verlag 1934.

"In experimental philosophy we are to look upon propositions inferred by general induction from phenomena as accurately or very nearly true, notwithstanding any contrary hypothesis that may be imagined, till such time as other phenomena occur, by which they may either be made more accurate or liable to exceptions."[81]

It's remarkable that Newton anticipated the possibility that his own theory may be liable to exceptions, which demonstrates a truly scientific disposition; Newton regarded his own theory as correct only insofar as it was supported by observation and measurement.

A Dramatic Difference

Despite the common theoretical ground between Newton and Einstein, the difference between them was not only dramatic; it was *melodramatic*. For example, one outcome of relativity is that a twin can fly off at the speed of light and return to find the other twin older. What an astonishing thing to say! It was as though a science fiction novel had been brought to life. Few events in human history can rival this fantastic level of drama.

However, if we were un-dramatic creatures of pure intelligence, the difference between Newton and Einstein would not seem so great. For example, with regard to time and space, we would merely conclude that the primary difference between Newton and Einstein is that Newton's *non-observable* absolute

---

[81] Isaac Newton, *The Mathematical Principles of Natural Philosophy*, from the section entitled "Rules of Reasoning in Philosophy," Rule IV.

time and space once *assumed* to be invariant and unconnected turned out upon observation to be connected and vary with respect to velocity. The *dramatic* difference between Newton and Einstein was far greater than the *scientific* difference. In fact, the drama of relativity was so great that the general public took and unprecedented interest in it, and non-scientific people from every walk of life seemed to have an opinion about the theory of relativity. Einstein referred to this phenomenon as the "relativity circus."[82]

Summary

Kuhn presented the shift between Newton and Einstein as a *particularly clear* example of how the conceptual network of one scientific theory is replaced by another, but the goal of Newton and Einstein were the same—to correctly understand natural phenomena as disclosed through the data of sense experience. In both cases the core scientific method was the same:

(1) the observation of data.
(2) insight into data.
(3) the formulation of the insight or set of insights.
(4) the verification of the formulation.[83]

Conceptual Similarities

Both systems regard the universe as a mathematical structure composed of atoms where physical events are

---
[82] Einstein, Archive 11-500.
[83] Bernard Lonergan, *Insight*, page 102.

absolutely subject to the law of cause and effect and the laws of nature are expressed as differential equations.

Both understand motion in terms of a coordinate system consisting of four numbers—three spatial coordinates and a time coordinate. With regard to the use of geometry, Euclid's absolute geometry is common to Newtonian physics and relativity, and the distance calculation in both systems involves the Pythagorean Theorem.

Common to both systems is the core notion of relativity, the principle of inertia, the principle of conservation, the notion of mass as the degree to which bodies resist acceleration, and the notion of energy as the ability to do work. In both systems: gravitation is universal; force is proportional to the change in momentum; inertial mass is equivalent to gravitational mass; and linear momentum, angular momentum and energy are conserved quantities.

Conceptual Differences

The two primary differences are:

(1) Time, space and mass, once thought by Newton to be unconnected and unchanging, were discovered to be connected and variable.
(2) The link between mass and energy was unknown to Newtonian science.

One might suppose the change from an absolute conception of time and space to a relative one is an example of a conceptual paradigm shift, but Newton's conception of time and space were, by definition, *non-observable* and therefore, *less scientific* than the notions of time and space that appear in the theory of relativity which *are* observable; consequently, the change in these concepts does not qualify as one of Kuhn's paradigm shifts.

Development

The theory of relativity integrated several core Newtonian concepts. This was achieved by extending the ancient principle of relativity discovered by Galileo to include the constant speed of light, which led to discovering a connection between the three dimensions of space and the dimension of time as well as the link between the core notion of mass and the core notion of energy, which constitutes three major developments:

(1) The principle of relativity was extended to include the constancy of light.
(2) The three dimensions of space were extended to include the dimension of time.
(3) The separate conservation laws of energy and mass were integrated into a single conservation law of mass-energy.

By omitting all scientific agreement between Newton and Einstein and by only calling attention to the *dramatic* differences between them, Kuhn exaggerated the *scientific* difference between Einstein and Newton and thereby gave the false impression that these two theories share no essential common ground.

Loss of Confidence in Reason

All is Dark

By distorting the rational nature of science, Kuhn contributed toward raising doubts about effectiveness of reason. The psychological story of how this came to pass begins just prior to the Enlightenment (roughly 1680–1780), when a scientific understanding of the physical world was generally shrouded in a mist of legend and superstition.

All is Light

Then Isaac Newton published *The Mathematical Principles of Natural Philosophy,* which demonstrated beyond doubt that the physical universe operates according to precise mathematical laws that could be reached by a rational inquiry. The transition was more than dramatic, as this quote by Alexander Pope (1688–1744) illustrates:

> "Nature and nature's laws lay hid by night,
> God said, 'Let there be Newton,' and all was light."

The world moved suddenly and unexpectedly from a shadowy understanding of the natural world to knowledge of exact scientific laws that could predict with astonishing precision the movements of all known heavenly bodies (including the tides of the earth).

Moreover, Newton's scientific knowledge did not reside in some ivory tower of academia but became the foundation of an immensely useful technology that had the power to transform daily life. The success of Newtonian science demonstrated that human beings could penetrate the secrets of nature and discover a hidden order that lies beyond the reach of direct sense experience but within the reach of rational inquiry, giving birth to a supreme confidence in reason.

Now consider a dramatic series of events that contributed toward a loss of confidence in reason since the Enlightenment. As we shall see, the publication of *The Structure of Scientific Revolutions* was one of three major cultural events that had a negative influence on the perception of human rationality:

(1) The Apparent Fall of Newtonian Physics.
(2) The Apparent Randomness of Natural Phenomena.
(3) The Publication of *The Structure of Scientific Revolutions*.

The Apparent Fall of Newtonian Physics

The success of Newtonian science signified a sweeping expansion of the power and effectiveness of human rationality—a confidence that soared for more than two hundred years until the early part of the 20th century. Then, according the Australian Philosopher, David Stove, something unthinkable happened: Einstein published the Special and General Theory of Relativity, apparently proving that Newton was wrong[84] and news that Newtonian physics, once thought to be incontestable and immortal, had collapsed profoundly shook our confidence in scientific claims and resulted in what Stove diagnosed as a

---

[84] David Stove, *Scientific Irrationalism Origins of a Post Modern Cult*, page 101.

"modern hyper-sensitivity about the possibility of scientific claims being wrong":

> "...it is this proposition, that any scientific theory, despite all the possible evidence for it, *might* be false: a proposition loudly announced by the fall of Newtonian physics; amplified ever since by morbidly sensitive philosophic ears; endlessly reapplied and reworded; insisted on to the exclusion of every other logical truth about science ... it is this proposition, so treated, which may be said to *be* the recent irrationalist philosophy of science."[85]

This loss of confidence in the effectiveness of scientific reason was unwarranted, because as we argued earlier, Newton was not fundamentally mistaken. To claim that he was simply wrong, as Kuhn claimed, is ignorant or misleading. For Newton did not say one single thing; he produced a rich body of scientific literature filled with an interconnected set of relevant insights into the natural world, much of which is still true today—a view that Einstein fully endorsed.[86]

Although Newton's fall was largely apparent, the psychological consequences were no less powerful. The perception that Newton was wrong introduced a profound uncertainty with regard to scientific claims.

---

[85] David Stove, *Scientific Irrationalism Origins of a Post Modern Cult*, page 189.
[86] From an article entitled "What is the Theory of Relativity?" *London Times*, November 28, 1919.

The Apparent Randomness of Natural Phenomena

If the apparent fall of Newtonian mechanics raised fundamental doubts about the strength and certainty of scientific reasoning, then the rise of quantum mechanics raised even deeper questions about the very rationality and comprehensibility of the universe itself.

The startling nature of quantum mechanics announced itself in the early part of the 20th century when a scientist by the name of Heisenberg discovered a strange and perplexing restriction in the attempt to measure the exact position and velocity of subatomic particles—a restriction commonly known as the "uncertainty principle." which may be explained in the following way: the more accurately one measures the position of a subatomic particle, the less accurately one can measure the velocity, and vice versa. The upshot of this discovery is that it is theoretically impossible to know the precise position and velocity of a subatomic particle at a given instant.

It is important to understand that prior to quantum mechanics it was considered at least *theoretically* possible to know the exact position and velocity of subatomic particles. One might, for example, possess a crude measuring device, but hypothetically one could measure these quantities precisely, provided that a way of obtaining such exact measurements was available. The discovery of the Heisenberg uncertainty principle proved that it is theoretically impossible to know the exact position and velocity of subatomic particles because the more accurately one property can be known, the less accurately the other property can be known.

Why can't we know the exact position and velocity of subatomic particle? This is one of the fundamental questions of quantum mechanics, and attempts to answer it have resulted in a proliferation of theoretical models, none of which can be fully substantiated at the present time. However, the commonly accepted theoretical model of quantum mechanics is an interpretation known as the "Copenhagen Interpretation," which states among other things that the act of measuring subatomic particles introduces a fundamental element of intrinsic randomness. The consequences of this view are staggering and carry profound implications. However, before we explore these implications let's be very clear about the exact meaning of "randomness" since there are two possible kinds: true and apparent.

The Difference between True and Apparent Randomness

Apparent randomness is the case where things appear to behave randomly but are absolutely determined. For instance, marbles spilling out onto a surface may *appear* to be moving randomly in all directions, but the exact movement of each marble is precisely *determined* by physical causes.

A truly random event is one that occurs for no reason at all. For instance, suppose a very thin domino stands upright on a perfectly flat surface on a perfectly windless day. Now suppose that it falls over for absolutely no reason at all. If this sounds strange, it is because it *is* strange, but this is exactly the sort of randomness that is commonly believed to take place in the subatomic realm. Thus, according the commonly accepted explanation, within certain limits, there is no particular reason why a particle appears at one position rather than another. But to propose that some things happen without reason explicitly denies the rational character of this universe - to be without reason is the very essence of irrationality.

Randomness Can Never Be Experimentally Verified

The apparently random behavior of subatomic particles is just as unverifiable as Newton's imaginary absolute time and space. The probability equations of quantum mechanics do not verify that subatomic events contain a necessary element of randomness; they only verify statistical relations between measurements. The failure to discover a cause for the behavior of subatomic particles does not prove that such behavior is uncaused; it is always possible to discover an explanation that is unexpected and surprising. For instance, being a magician is the subtle art of making things appear to happen without reason. If this is done by an expert, it can make people believe in magic. It may be that Mother Nature is a sufficiently subtle magician that she can even make a scientist believe in magic. Of course, the opposite argument can be made, which is that we cannot prove that there are no random events. Even if we discovered all the causes of subatomic events, it is still logically possible that some events occur without reason. We simply do not know. However, the general acceptance that some events in this universe occur without reason had deep psychological consequences.

A Deeper Problem with Assuming Randomness

Randomness is not an explanation; it is the lack of an explanation. It is not an insight into phenomenon; it is an absence of insight. Randomness is simply a name assigned to behavior for which we have not discerned a pattern or cause. The *assumption* that some events are uncaused is not only unverifiable, but also in a deeper sense is fundamentally opposed to the whole nature and essence of scientific inquiry.

All genuine scientific investigation begins with a desire to know. Thus science by its very nature is a kind of adventure, an exploration of possibility—the possibility of understanding this world. Moreover, such exploration is undertaken without any guarantee or promise of success. In this sense, there is something optimistic and enterprising about the scientific spirit. Science is *rational* inquiry insofar as it is a search for the causes of physical phenomena. To assume the *unverifiable* proposition that certain events happen without causation defeats the whole point of rational inquiry.

The rational nature of science is grounded in the law of cause effect. If there were no causal relations, then science would be impossible. One could not build reliable cars, radios and satellite stations if physical events in the universe were governed by pure chance. This is also true of Quantum Mechanics which, like the rest of modern science, ultimately rests on the principle of causality - If it did not, we would not be able to build machines such as high precision lasers based on Quantum Mechanics though this point is seldom remembered. For just as the dramatic discovery of Relativity made Newton appear to be wrong, the dramatic discovery of the uncertainty relations made the principle of causality seem mistaken.

Quantum Mechanical relations are based on a type of causality. An ordinary casual relation might be stated like this: Event A necessarily determines one and only one result which we shall call Event B: A→B. For example suppose Event A is a rock being thrown at a window and Event B is the window breaking. In Quantum Mechanics this relation is not abandoned but *modified*, thus, Instead of an event determining one and only one result, an event determines a *range* of possible results that *must* fall with certain boundaries. For example, suppose a coin is tossed and the result of the toss will be heads or tails. Let us refer to the toss of the coin as Event A. Let H designate the side of the coin known as heads. Let T designate the side of the coin known as tails. So that we can say that Event A *necessarily determines* that the result will be H or T: A → (H or T). Since, in Quantum Mechanics, no one has been able to discover why the result within the specified range is one value and not another (in this case heads or tails), it has been assumed the result is random and that no explanation exists.

Now the point of this section is not to offer a well defended position on the nature of the uncertainty relations. Rather, the point here is to draw attention to the psychological consequences of quantum mechanics which is this: The widespread acceptance among scientists that subatomic events can happen without reason ultimately raised doubts about the rational nature of this universe and consequently deepened our doubts about effectiveness of rational inquiry.

The Publication of *The Structure of Scientific Revolutions*

Let us approach the psychological influence of Thomas Kuhn by briefly reviewing the last two sections. The birth of Newtonian physics (1792) gave rise to an unbridled confidence that rational scientific inquiry can yield a timeless knowledge of physical reality in the form of empirical equations. This supreme confidence lasted for more than two hundred years until the melodramatic discovery of relativity (1927), which, like an optical illusion, made Newton appear to be wrong and deeply shook our confidence in the power of rational thought. This confidence was further shaken by widespread acceptance of the unverifiable hypothesis that certain sub-atomic events occur for no reason at all.

With the apparent fall of Newtonian mechanics raising doubts about the certainty of scientific claims and the possibility that things can happen without reason, it started to seem as though the power and effectiveness of rationality had been greatly over estimated. It was in this atmosphere of doubt and confusion that Kuhn published *The Structure of Scientific Revolutions* (1962), which offered a philosophical justification for the widespread doubts about the certainty and rationality of scientific knowledge—a task aided by the fact that Kuhn wrote in the 60's, a time when challenges to established authority (including scientific authority) could count on enthusiastic support.

According to Kuhn, scientific inquiry does not lead to a universal and unchanging knowledge of nature, as had been believed for more than two centuries. On the contrary, Kuhn argued that scientific knowledge is periodically rebuilt from the foundations. His clearest example from history was nothing other than the apparent fall of Newtonian mechanics.[87] Moreover, the impression than Newtonian physics had collapsed was so palpable that Kuhn scarcely needed to mention it, which perhaps explains why his remarks on the subject are confined to a brief enumeration of the most striking and dramatic differences between Newton and Einstein.[88]

    I would like to conclude this work by expressing myself poetically. Science, at the first glance, can seem dry and mundane, but nonetheless possesses a power and a beauty that, once perceived, is a source of endless wonder and delight. Science is a brilliant sun that can light up the nature of things—it radiates the light of reason. If the rise of scientific irrationality continues to eclipse that light, we may need a New Age of Reason.

---

[87] Thomas Kuhn, *The Structure of Scientific Revolutions*, page 102.
[88] Thomas Kuhn, *The Structure of Scientific Revolutions*, page 98–102.

Bibliography

The Structure of Scientific Revolutions, Thomas Kuhn, 1996, Third Edition, The University of Chicago Press, ISBN: 0-226-45808.

The Principia, Isaac Newton, 1995, Prometheus Books, ISBN: 13: 978-0-87975-980-3.

Great Physicists, William H. Cropper, 2001, Oxford University press, ISBN: 0-195-13748-5.

Newton Versus Relativity, Jean-Michel Rocard, 1992, Vintage Press, ISBN: 0-533-09637-5.

The Scientists, John Gribbin, 2004, Random House Trade Paperback, ISBN: 0-8129-6788-7.

Ideas and Opinions, Albert Einstein, 1982, Crown Publishers, ISBN: 0-517-88440-2.

The Expanded Quotable Einstein, Alice Calaprice, 2000, Princeton University Press,
 ISBN: 0-691-07021-0.

On the Shoulders of Giants, Stephen Hawking, Running Press books Publishers, ISBN: 0-7624-1348-4.

The Cambridge Companion to Newton, Bernard Cohen and George E. Smith, 2002, Cambridge University Press, ISBN: 0-52165696-6.

Insight, Bernard Lonergan, Longmans, Green & Co., London, 1957, ISBN: 0-8020-3454-3.

The Feyman Lectures on Physics, 1977, Addison-Wesley Publishing Company, ISBN: 0-201-02116-1.

The Character of Physical Law, Richard Feyman, 1967, MIT Press, ISBN: 0-262-06016-7.

Mathematics and the physical world, Morris Cline, 1981, Dover Publications, Inc., New York, ISBN: 0-486-24104-1.

The meaning of relativity, Albert Einstein, 1974, Fourth Princeton Paperback Printing, ISBN: 0-691-02352-2.

The Portable Enlightenment Reader, Isaac Kramnick, 1995, Penguin Books, ISBN: 0-14-02-4566 -9.

Homage to Galileo, Morton F Kaplan, 1964, The M.I.T Press ISBN: 65-27973.

Galileo Studies, Stillman Drake, 1970, The University of Michigan Press ISBN: 0-472-8283-3.

The Complete Idiots Guide to Understanding Einstein, Gary F. Moring, 2004, Alpha,
ISBN: 1-59257-185-9.

Albert Einstein and the Frontiers of Physics, Jeremy Bernstein, 1996, Oxford University Press, ISBN: 0-19-512029-9.

Atom Journey Across the Subatomic Cosmos, Issac Asimov, 1992, Truman Talley Books, ISBN: 0-452-26834-6.

Return to Reason, Stephen Toulmin, 2001, Harvard University Press, ISBN: 0-674-00495-7.

Galileo, Stillman Drake, Hill and Wang, New York, ISBN: 0-8090-4850-7.

A Brief History of Time, Stephen Hawking, Bantam Books, 1996, ISBN: 0-553-10953-7.

Scientific Irrationalism Origins of a Post Modern Cult, David Stove, Transaction Publishers 2001, ISBN: 0-7658-0063-2.

The Essential Tension, Thomas Kuhn, The University of Chicago Press, 1977, ISBN: 0-226-45806-7.

Galileo: Pioneer Scientist, Stillman Drake, University of Toronto Press, 1990, ISBN: 0-8020-2725-3.

Galileo's Intellectual Revolution, Middle Period 1610–1632, William R. Sheaw, Science History Publications, 1972, ISBN 0-88202-006-4.

www.ingramcontent.com/pod-product-compliance
Lightning Source LLC
Chambersburg PA
CBHW051454290426
44109CB00016B/1746